Reducing Greenhouse Gas Emissions and Improving Air Quality

Reducing Greenhouse Gas Emissions and Improving Air Quality

Two Interrelated Global Challenges

Larry E. Erickson and Gary Brase

CRC Press is an imprint of the
Taylor & Francis Group, an informa business

CRC Press
Taylor & Francis Group
6000 Broken Sound Parkway NW, Suite 300
Boca Raton, FL 33487-2742

Printed on acid-free paper

International Standard Book Number-13 978-0-8153-6127-5 (Hardback)
International Standard Book Number-13 978-0-367-40875-6 (Paperback)

Library of Congress Cataloging-in-Publication Data
Names: Erickson, L. E. (Larry Eugene), 1938– author. | Brase, Gary, author.
Title: Reducing greenhouse gas emissions and improving air quality : two interrelated global challenges / authored by L Larry E. Erickson and Gary Brase.
Description: Boca Raton : CRC Press, 2020. | Includes bibliographical references and index. | Summary: "Our lives are impacted by energy, transportation, air quality, and climate change. This book will help readers understand the transitions that happen at the nexus of these areas, as they relate to reducing greenhouse gas emissions and improving air quality, and the benefits of these transitions for electric vehicles, solar and wind energy for electricity generation, battery developments, smart grids and electric power management, and progress in the electrification of agricultural technology. It aims to provide the latest information in the context of the United Nations sustainable development goals and the Paris Agreement on Climate Change"–Provided by publisher.
Identifiers: LCCN 2019026551 | ISBN 9780815361275 (hardback : acid-free paper) | ISBN 9781351116589 (ebook)
Subjects: LCSH: Greenhouse gas mitigation. | Air quality management. | Sustainable development. | Energy conservation. | Climatic changes.
Classification: LCC TD885.5.G73 E75 2020 | DDC 628.5/32–dc23
LC record available at https://lccn.loc.gov/2019026551

Visit the Taylor & Francis Web site at
www.taylorandfrancis.com

and the CRC Press Web site at
www.crcpress.com

Contents

Preface

This book is for all readers who want to know more about not only air quality and greenhouse gas emissions, key factors contributing to climate change, but also the rapidly changing technologies and social factors that are being used to address these issues. These technologies include the electrification of transportation, renewable energy sources, battery storage of electricity, and the development and management of a smart grid. The social factors include government policies towards these technologies and how they are used, economic considerations, and human decision making about how air quality and emissions are managed. Finally, there are some specific illustrations of contexts (agriculture) and locations that are of particular interest.

Our interest in all of this began expanding in 2009, in the context of a National Science Foundation (NSF) supported Research Experience for Undergraduates program: 'Earth, Wind, and Fire: Sustainable Energy for the 21st Century'. Some of the research within that program started looking at adding solar panels to parking lots in order both to generate electricity and to have shaded parking. Many of the undergraduate students were also interested in reducing greenhouse gas emissions, sustainable energy, and climate change – it was clear that there were many directions we could go with this work. Renewable energy has also been a popular research area because of its importance to society. Common Sustainable Development Goals include a transition to renewable energy, the generation of electricity without carbon emissions, and the electrification of transportation in order to have cities with zero-emission vehicles.

Since 2009, there has been great progress in the development of solar panels, batteries, and electric vehicles. Our earlier book, *Solar Powered Charging Infrastructure for Electric Vehicles: A Sustainable Development* (Erickson et al., 2017a), provided information on developments related to the generation of electricity with solar photovoltaics (PV), electric vehicle charging infrastructure, batteries, and electric vehicles. One of the very interesting aspects of these developments is that they can interact when they are combined to provide benefits even greater than one might expect from just the individual innovations.

Our earlier book includes a chapter on urban air quality because of the health benefits of the transition to renewable energy and electric vehicles. Because of our subsequent work on a US Environmental Protection Agency project (RD83618201), in which we worked with four intercity communities in Chicago on measuring air quality, we became immersed in the literature on air quality and health (Erickson, 2017; Erickson and Jennings, 2017; Erickson et al., 2017b). This work resulted in a better understanding of air quality and health; actions to improve air quality in the pursuit of improving health can have great value with respect to reducing greenhouse gas emissions.

We realized that there is a need to help people understand, with this new book, all the ways that the two topics of reducing greenhouse gas emissions and improving air quality are related. Indeed, they can interact and feed off of each other much like electric vehicles and solar power charging can interact to provide multiplicative benefits.

Since air quality, electricity, and transportation impact everyone, this book addresses important topics that people need to understand. At this time, in 2019, there has been great progress in the transition to electricity from renewable sources and the electrification of transportation. The prices associated with solar PV, wind energy, batteries, and electric vehicles are much better than in past years, and they are improving.

Additionally, it is important for everyone to realize that the world's atmosphere is a common resource. All people consume air and depend on good air quality for their health. Air pollution is associated with more than $4 trillion in social costs each year, including many people who die prematurely. We now have good options to improve air quality. We can add solar panels to our home and generate some of the electricity that we use. We can add insulation and reduce energy use for heating and cooling.

We all have many choices with regard to taking action to reduce greenhouse gas emissions and improve air quality. It is possible to do one or two small things each day to improve air quality and reduce emissions, such as walking to work or going by bicycle. It is also possible to do larger things to improve air quality and reduce emissions, such as shifting to electric transportation, generating renewable power, and enacting government policy changes. You can read this book and apply some of the ideas in your daily life.

References

Erickson, L.E. (2017). Reducing greenhouse gas emissions and improving air quality: Two global challenges. *Environmental Progress and Sustainable Energy* 36: 982–988.

Erickson, L.E. and Jennings, M. (2017). Energy, transportation, air quality, climate change, health nexus: Sustainable energy is good for our health. *AIMS Public Health* 4: 47–61.

Erickson, L.S., Robinson, J., Brase, G., and Cutsor, J. (2017a). *Solar Powered Charging Infrastructure for Electric Vehicles: A Sustainable Development*. CRC Press, Boca Raton, Florida.

Erickson, L.E., Griswold, W., Maghirang, R.G., and Urbaszewski, B.P. (2017b). Air quality, health, and community action. *Journal of Environmental Protection* 8: 1057–1074.

Acknowledgments

Many people have contributed to our understanding of an array of issues that form the foundations of this book. Chief among these are the interactions with faculty, staff, and students at Kansas State University through the NSF Research Experiences for Undergraduates (REU) program, CHE 670 Sustainability Seminar, CHE 750 Air Quality Seminar, the annual Dialog on Sustainability, the Consortium for Environmental Stewardship and Sustainability (CESAS), chemical engineering design project activities, and other events.

Martin Abraham had the vision to encourage review articles in *Environmental Progress and Sustainable Energy*, and we thank him for the invitation to prepare the review 'Reducing Greenhouse Gas Emissions and Improving Air Quality: Two Global Challenges'. This paper resulted in an invitation to expand the subject matter into this book.

Many of the ideas that are included in this book are related to the research that was done under Assistance Agreement RD83618201 awarded by the US Environmental Protection Agency (EPA) to Kansas State University. This book has not been formally reviewed by the EPA. The views expressed in this book are solely those of the authors and do not necessarily reflect those of the Agency. The EPA does not endorse any products or services mentioned in this publication.

Many good ideas came from dialogue with the many people who participated in public education events and environmental organizations. Thank you in particular to members and friends of the Climate and Energy Project, Flint Hills Renewable Energy and Efficiency Cooperative, Kansas Natural Resource Council, Lutherans Restoring Creation, Kansas Interfaith Action, Clean Air Now, Sierra Club, Metropolitan Energy Center, and Kansas Rural Center.

We thank Proterra, Inc. for permission to use the photo on the cover of the book. This 35-foot Catalyst Bus is one of the electric models that Proterra manufactures and sells. For more information, see www.proterra.com.

We express special thanks to Danita Deters for her indefatigable help with the manuscript and other aspects of developing the book. Sheree Walsh has been very helpful in hosting many activities that are related to the topics of the book, and we profusely thank her for her work.

And, last on this list but absolutely not least, we thank Irma Britton for all the many ways she has helped with this book, and we thank Sandra Brase and Laurel Erickson for their support and encouragement.

Larry E. Erickson
Gary Brase

The Authors

Larry E. Erickson is Professor of Chemical Engineering at Kansas State University. His research interests include environmental engineering, air quality, and sustainable development. His teaching includes an air quality seminar and a sustainability seminar. He helps to provide leadership for the Consortium for Environmental Stewardship and Sustainability and the annual Dialog on Sustainability. His degrees are in chemical engineering from K-State.

Gary Brase is a professor and the Graduate Program Director in the Department of Psychological Sciences at Kansas State University. Dr. Brase studies complex human decision making using cognitive and evolutionary theories, including topics such as medical decisions, decisions about sustainability issues, relationship decision making, and reasoning about social rules. He has over 75 journal articles and other publications, and over 100 research presentations. Dr. Brase received his PhD from the University of California, Santa Barbara in 1997 and has been at K-State since 2007.

Contributors

Ronaldo G. Maghirang is Professor of Biological and Agricultural Engineering, and Associate Dean for research and graduate programs at Kansas State University. His research is on air quality.

Kate Preston is an undergraduate student in chemical engineering at Kansas State University who helped with the chapter on batteries.

1

Introduction

Abstract

The 2015 Paris Agreement on Climate Change, supported by the major countries of the world, sets a global path towards many outcomes. Reducing greenhouse gas emissions and improving air quality are two important goals that have strong support from many people. The efforts to reduce greenhouse gas emissions address a global climate challenge and will also help to improve air quality overall and particularly in local urban communities. Additionally, there are huge economic benefits associated with these goals, exceeding $3 trillion/year. Several transitions to reach these goals have already started, including increases in wind and solar energy generation and improved storage technologies. This new path includes a number of complex challenges, though, in terms of managing the interactions of electricity generation, the smart grid, energy storage (e.g. in electric vehicles), peak power management, air quality, and health. This book addresses a number of these complex challenges, outlining how they can be dealt with in ways that help to advance the Sustainable Development Goals.

1.1 Introduction

No one likes to walk an unfamiliar path in the dark. The purpose of this book is to shed some light on the still dimly lit path we need to take with regard to reducing greenhouse gas emissions and improving global air quality. Why do we need to take this new path? Why not continue down the path we are on now? Briefly, the path we are on now – relying on fossil-based fuels (oil, gas, and coal) – is clearly and increasingly a very wrong path. The fossil fuels path will inevitably be a dead end because it rests on non-renewable resources; we will someday run out of oil, then gas, then coal to extract. In the meantime, reliance on oil (e.g. from the Middle East) and gas (e.g. from Russia) continues to create political conflicts and entanglements for many countries. And all the while these fuels generate massive amounts of pollution as they are extracted, refined, and then burned. Finally, these fuels are major contributors to global climate change. The old path is a road to ruin.

But the new path is… new, and that can be scary. Better illumination of this path will help to bring many people along, and the goal of this book is to help provide some of that light.

Like any new direction, there are a number of different aspects to explore. Erickson et al. (2017) spent an entire book looking at just the effects of more electric vehicles, greater solar power infrastructure, and the interactions of these two trends. The current book includes these parts of the new path, but also expands to focus on air quality, greenhouse gas emissions, climate change, and the interrelationships between these issues.

The first steps towards this new path were set out by an international accord often referred to as the Paris Agreement (see Chapter 2). On December 12, 2015, the Paris Agreement on Climate Change was adopted by the Parties to the United Nations Framework Convention on Climate Change (UNFCCC, 2015). This agreement has a goal to reduce greenhouse gas emissions until carbon dioxide concentrations in the atmosphere stop increasing (Erickson et al., 2017; UNFCCC, 2015). The Paris Agreement came into force on November 4, 2016 after more than 55 countries that emit more than 55% of greenhouse gases ratified the agreement (Erickson, 2017). In November 2017, the UN Conference of the Parties was held in Bonn, Germany, with many meetings to address implementation issues; however, the significant action is happening in the member countries, where efforts are in progress to reduce greenhouse gas emissions. China is moving forward with actions to increase the installation of solar panels for electricity generation, and sales of electric vehicles are increasing rapidly. In September 2017, plug-in electric and hybrid vehicle sales were about 59,000 in China compared to about 21,000 in the USA and about 123,000 in the world (Bohlsen, 2017; Cole, 2017). In 2017, sales in China exceeded 570,000 vehicles, and at the end of 2017, more than 3 million plug-in vehicles were in service (McCarthy, 2018). A number of major car manufacturers (Volkswagen, Mercedes-Benz) have plans to eventually shift all of their new production to electric vehicles.

1.2 Greenhouse Gas Emissions

The Paris Agreement focuses on greenhouse gas emissions for several reasons: They can be defined and measured fairly clearly, and they are at the nexus of the fundamental concerns that motivate the Agreement. The goal to reduce greenhouse gas emissions by 80% by 2050 has been proposed and it is considered to be a reasonable objective. Accomplishing that goal, though, will involve an array of short-, medium-, and long-range activities. For example, Erickson et al. (2017) outlined a number of issues related to the electrification of transportation (i.e. adopting electric vehicles), the transition to renewable energy and smart grids, and the need for infrastructure for electric vehicles. These issues are interrelated because the electrical grid is intimately connected to the powering of electric vehicles.

The topics covered by Erickson et al. (2017) were just about these particular parts of this new pathway (electric vehicles, renewable energy, and the electrical grid). There is a much wider view of the situation that this book addresses. This wider view includes the immediate crisis that motivates the focus on greenhouse gas emissions – air pollution – and the longer-term crisis from those same emissions – climate change.

1.3 Air Pollution

Greenhouse gas emissions are a major part of general air pollution. Although air pollution includes many other elements, reductions in greenhouse gas emissions will tend to cut down several other contributors to air pollution as a byproduct. Air pollution is the greatest environmental risk to health (see Chapter 3). In 2012, according to the World Health Organization (WHO), air pollution contributed to one out of every nine deaths (WHO, 2016). Combustion emissions, their health effects, and how to reduce emissions are an important part of this book. Air pollution in large cities and metro areas is a global health concern, with social costs of more than $3 trillion every year and more than 3 million people dying prematurely each year because of outdoor air pollution (Lelieveld et al., 2015).

Emissions from transportation and coal-fired electricity generation are major sources of air pollution (including a lot of greenhouse gas emissions). The electrification of transportation (Chapter 4) and the transition to renewable energy for the generation of electricity (Chapters 5–9) thus have two important benefits: Improved air quality and reduced greenhouse gas emissions. Indeed, the intertwining of these outcomes can help to make sense of some otherwise puzzling behaviors. For example, one of the reasons China is electrifying transportation is because air quality is very poor in many of the larger cities, even as it uses coal power plants to generate a lot of the required electricity. The next step (for China, and for the world) is to shift the electrification of cities to carbon-free energy generation from sources such as wind and solar. Transitions of both electricity generation and the transportation sector will improve air quality greatly (Stewart et al., 2018).

1.4 Climate Change

The other reason to focus on greenhouse gas emissions is that they are a major contributor to global climate change. Climate change is an issue on a much larger scale and a much longer time frame than air pollution, but it also is an even more dire crisis than air pollution. Indeed, the scale of climate change (both in size and time) is such that it makes putting exact numbers on the climate change situation difficult. What we do know is that those numbers are immense and steadily getting even bigger. The global costs of climate change have been increasing because of the increase in concentrations of greenhouse gases in the atmosphere. In 2017, there were wildfires in many midwestern and western parts of the United States, and these cost many billions of dollars. But the greatest costs from climate change in 2017 (in the United States, at least) were because of hurricanes in Texas, Florida, and Puerto Rico. Rainfall associated with Hurricane Harvey resulted in flooding and damage to property. In Puerto Rico, there was major damage to property, and electrical service was destroyed in many locations. More forest fires in 2018, just in California, have again cost billions of dollars and the loss of many human lives. In 2019, there was flooding in Nebraska that cost billions of dollars and impacted the lives of many people.

1.5 Economics

Reducing greenhouse gas emissions, air pollution, and the pace of climate change are all positive things that sound good in principle, but for many people it will have to make financial sense for them to take any action. Economics and prices influence decisions that are made at nearly all levels, from individual decisions (e.g. whether to buy an electric vehicle; Brase, 2018) to international accords. If we think of this as a path – one that we are trying to light more clearly – the economics of the situation (see Chapter 11) determine whether we are trying to walk uphill or downhill. Fortunately, the economics surrounding these issues are pretty favorable and steadily improving (i.e. walking downhill).

The reductions in costs for solar panels, wind-generated electricity, and batteries for energy storage in electric vehicles are already having major impacts. The costs of wind power, solar-generated electricity, electric vehicles, and batteries are falling because of research progress, production experience, competition, and the magnitude of commercial activity (Erickson and Jennings, 2017). The transition to electric vehicles has the potential to be very beneficial for many people because of reduced costs and improved air quality. As costs continue to fall, electric vehicle sales are increasing and plug-in vehicles are becoming more popular. In 2017, global sales of plug-in all-electric and hybrid vehicles exceeded 1 million passenger cars for the first time. This was followed by more than 2 million new electric vehicles (EVs) delivered in 2018 (Irle, 2019; Kane, 2019). Policy incentives, such as the plan by France to end sales of gasoline and diesel cars by 2040, have also had an impact (Ewing, 2017). London has daily fees that are larger for polluting vehicles to try to improve air quality (Erickson, 2017).

The great transitions to wind and solar electricity generation and to zero-emission vehicles have started (Chapters 12–14). The economic progress towards competitive, low-cost innovations is very important because many consumers are influenced by cost savings.

1.6 Complex Interactions

Each of the above topics – greenhouse gas emissions, air pollution, climate change, electrification of transportation, smart grids, and the economics of all these – are large and important topics. They are also all interrelated. This book addresses important interactions, including the energy, transportation, air quality, climate change, and health nexus (Erickson and Jennings, 2017); the renewable energy, electric vehicle, and smart grid nexus; and the Sustainable Development Goals, air quality, health, renewable energy, and electric vehicle nexus. Indeed, although each of these topics is plenty complicated on its own, there is also significant complexity in their interactions. For instance, wind- and solar-generated electricity must be managed by finding uses for the amount of electricity being generated, at the time that it is generated. One way to handle this interaction dynamic is by leveraging the possibility that electric vehicle batteries can be charged when power needs to be delivered, thus helping to balance supply and demand (Chapters 4 and 7). In many parts of the world, there are off-grid opportunities to generate electricity with solar panels and store it in batteries because prices are decreasing for both batteries and solar panels (Chapter 9).

Sector integration is one of the significant issues in urban sustainable development, and this can include efficient management of green buildings, transportation, water and waste management, and ecosystem care. This integration and coordination itself is a challenge. Good participatory governance in a multisector, multistakeholder environment is often a challenge for community leaders (Stewart et al., 2018). There is a challenge in the need to understand social, institutional, economic, physical, and technological limitations regarding the process of advancing sustainability (see Chapters 15 and 16). Nevertheless, it is crucial that we take on these challenges – and take them on now – because improved air quality is one of the most significant benefits of reducing combustion processes in large urban cities.

Any list of goals and interactions across these areas is necessarily going to be incomplete, but we can start with an initial summary. The transition to sustainable energy without carbon emissions and towards electric transportation has the following major challenges:

1. Developing a stable electrical grid powered by renewable energy that does not have carbon emissions.
2. Further developing electric vehicles so that most transportation is powered by electricity and other energy sources that do not produce greenhouse gases.
3. Developing an electric vehicle charging network such that there are charging stations in most places where they are needed, including along highways, in shopping malls, places of work, hotels and motels, restaurants, parking garages, airports, and libraries.
4. Modernization of the electrical grid with better communication to enable real-time prices or time-of-use prices to be used for the purchasing of electricity.
5. Integration of electric vehicle batteries into the electrical grid such that the charging of electric vehicle batteries is an integral part of managing the electrical grid.
6. Modernization of the electrical grid with agreements that allow the charging of electric vehicles to be managed by the grid manager.
7. Including the value of better health and improved air quality in the analysis of the costs and benefits of sustainable development projects that improve air quality and advance renewable energy and electric vehicles.

The efforts to reduce greenhouse gas emissions and improve air quality are closely related to many of the Sustainable Development Goals and the global challenge to improve the quality of life for all people.

1.7 Sustainable Development Goals

A set of Sustainable Development Goals was adopted in September 2015 at the United Nations in New York by the heads of state and government and high representatives. These new goals replaced the Millennium Development Goals, which have quietly been immensely successful. The Millennium Development Goals lifted more than 1 billion people out of extreme poverty

TABLE 1.1

Sustainable Development Goals.

1.	End poverty in all its forms everywhere
2.	End hunger, achieve food security and improved nutrition and promote sustainable agriculture
3.	Ensure healthy lives and promote well-being for all at all ages
4.	Ensure inclusive and equitable quality education and promote lifelong learning opportunities for all
5.	Achieve gender equality and empower all women and girls
6.	Ensure availability and sustainable management of water and sanitation for all
7.	Ensure access to affordable, reliable, sustainable and modern energy for all
8.	Promote sustained, inclusive and sustainable economic growth, full and productive employment and decent work for all
9.	Build resilient infrastructure, promote inclusive and sustainable industrialization and foster innovation
10.	Reduce inequality within and among countries
11.	Make cities and human settlements inclusive, safe, resilient and sustainable
12.	Ensure sustainable consumption and production patterns
13.	Take urgent action to combat climate change and its impacts
14.	Conserve and sustainably use the oceans, seas and marine resources for sustainable development
15.	Protect, restore and promote sustainable use of terrestrial ecosystems, sustainably manage forests, combat desertification and halt and reverse land degradation and halt biodiversity loss
16.	Promote peaceful and inclusive societies for sustainable development, provide access to justice for all and build effective, accountable and inclusive institutions at all levels
17.	Strengthen the means of implementation and revitalize the global partnership for sustainable development

Source: UN, 2015.

worldwide, reduced child mortality, and reduced the percentage of children not in school (UN, 2015).

The goals of this book are supportive of a number of the 2015 Sustainable Development Goals, which include economic, social, and environmental dimensions, as well as the global commitment that has been accepted by all countries. The Sustainable Development Goals are listed in Table 1.1, and a few of them are particularly worth noting for our current context. Goal 3 includes improving urban air quality because that contributes to healthy lives and well-being. Goal 3.9 includes reducing the number of deaths because of air pollution (UN, 2015). Goal 7 relates to developing sustainable energy and reducing greenhouse gas emissions. Goal 7.2 is to increase substantially the fraction of renewable energy in the global mix (UN, 2015). Goal 11 relates to making cities safe and sustainable. Goal 11.6 includes improving air quality (UN, 2015). This book will address urban topics related to health and the transformation to more sustainable transportation and energy. Goal 13 addresses climate change, which relates to the emphasis on reducing greenhouse gas emissions. Goal 13.2 is to integrate climate change measures into policies, strategies, and planning (UN, 2015). Goal 16 includes justice for all; one aspect of environmental justice is to have good air quality for all people. Goal 16.3 is to promote the rule of law and ensure access to justice. Goal 16.6 is to develop effective, accountable, and transparent institutions (UN, 2015). Goal 17 addresses implementation; one of the goals of this book is to address a number of sustainable development issues that will be beneficial to readers who are working on sustainable energy, electrification of transportation, modernization of the electrical grid, and improvement of air quality, which are important aspects for those involved in the global partnership for sustainable development.

The Sustainable Development Goals are for the time period from January 1, 2016 to 2030, with the hope that all people and nations will prosper and participate. Those involved in adopting these goals recognize that "climate change is one of the greatest challenges of our time" and that "sustainable urban development and management are crucial to the quality of life of people" (UN, 2015).

1.8 Linkage Between Reducing Greenhouse Gases and Improving Air Quality

Since climate-changing combustion of fossil fuels produces both greenhouse gases and other emissions that impact air quality, it is logical to address both climate change and air quality together. Policy decisions should consider the collective impacts on greenhouse gas emissions, climate change, and air quality (Melamed et al., 2016). The Paris Agreement on Climate Change has a focus on reducing greenhouse gas emissions; however, those working to implement policies to reduce emissions can also take into account the impact of their actions on urban air quality and the Sustainable Development Goals.

1.9 Ecological Sustainable Development

The importance of including social, environmental, and economic (triple bottom line) concerns in decision making will be considered in this book, because it is beneficial to consider health and well-being as well as ecological integrity in making decisions (Lederwasch and Mukheibir, 2013). Both the long-term and short-term impacts of decisions are important because new vehicles, power plants, and buildings often have long lives. Efforts to improve air quality and reduce greenhouse gas emissions will need to be continued over multiple generations.

References

Bohlsen, M. (2017). EV company news for the month of October 2017. *Seeking Alpha;* https://seeking alpha.com

Brase, G.L. (2018). What would it take to get you into an electric car? Consumer perceptions and decision making about electric vehicles. *The Journal of Psychology* 153: 214–236.

Cole, J. (2017). October 2017 plug-in electric vehicle sales report card. *Inside EVs;* https://insideevs.com

Erickson, L.E. (2017). Reducing greenhouse gas emissions and improving air quality: Two global challenges. *Environmental Progress and Sustainable Energy* 36: 982–988.

Erickson, L.E. and Jennings, M. (2017). Energy, transportation, air quality, climate change, health nexus: Sustainable energy is good for our health. *AIMS Public Health* 4: 47–61.

Erickson, L.E., Robinson, J., Brase, G., and Cutsor, J. (2017). *Solar Powered Charging Infrastructure for Electric Vehicles: A Sustainable Development*. CRC Press, Boca Raton, Florida.

Ewing, J. (2017, July 6). France plans to end sales of gas and diesel cars by 2040, *New York Times;* www.nytimes.com

Irle, R. (2019). Global EV sales for 2018 – Final results. *EV Volumes.com;* www.ev-volumes.com/

Kane, M. (2019). Amazing year, amazing growth and records. *Inside EVs;* insideevs.com/

Lederwasch, A. and Mukheibir, P. (2013). The triple bottom line and progress toward ecological sustainable development: Australia's coal mining industry as a case study. *Resources* 2: 26–38.

Lelieveld, J., Evans, J.S., Fnais, M. et al. (2015). The contribution of outdoor air pollution to premature mortality on a global scale. *Nature* 525: 367–371.

McCarthy, N. (2018, June 4). Electric car sales are surging in China. *Statista;* www.statista.com

Melamed, M.L., Schmale, J., and Schneidermesser, E.V. (2016). Sustainable policy – key considerations for policy and climate change. *Current Opinion in Environmental Sustainability* 23: 85–91.

Stewart, I.D., Kennedy, C.A., Facchini, A., and Mele, R. (2018). The electric city as a solution to sustainable urban development. *Journal of Urban Technology* 25: 1, 3–20.

UN (2015). *Transforming Our World: The 2030 Agenda for Sustainable Development*. United Nations, New York; see also *Sustainable Development Goals*. United Nations, New York.

UNFCCC (2015). Paris Agreement. *United Nations Framework Convention on Climate Change*. United Nations.

WHO (2016). *Ambient Air Pollution: A Global Assessment of Exposure and Burden of Disease*. World Health Organization, Geneva, Switzerland.

2

Paris Agreement on Climate Change

Abstract

There has been significant progress since 2015, when the Paris Agreement on Climate Change was adopted. Consistent with the goals of the Agreement, many coal-fired plants have been retired, as the costs of electricity from solar photovoltaic systems and wind power have decreased to being less expensive than electricity from coal. Most of the new electricity-generating capacity that is being developed and put into service is from renewable wind and solar systems. There is still far to go, however. Trillions of dollars can be saved by reducing greenhouse gas emissions and reaching the state where carbon dioxide concentrations in the atmosphere are no longer increasing. There are also many benefits associated with the transition to zero carbon emissions and steady state concentrations of greenhouse gases. Advances in science and technology that reduce the costs of electric vehicles and battery storage contribute to reaching the goals of the Paris Agreement on Climate Change.

2.1 Introduction

Climate change is a fact of life in today's world, but what are the implications of those changes? This chapter reviews what climate change means for people right now, what it will mean for future generations, and what needs to be done to avert those consequences. It is very important to understand how the climate is changing and how it is affecting life on Earth. Although it is a very significant future concern, present events influenced by climate change are already of great importance. Human suffering is already being inflicted by the influences of climate change on hurricanes, wildfires, floods, droughts, sea level rise, and increased temperatures. The problems posed by climate change are well established, but the actions taken so far to address these problems are inadequate.

2.2 Goals of the Paris Agreement

A key benchmark for understanding what needs to be done to manage the effects of climate change is the United Nations Paris Agreement on Climate Change, which was adopted on December 12, 2015 by the Parties of the United Nations Framework Convention on Climate Change (UNFCCC, 2015) and came into force on November 4, 2016 after appropriate ratification. The Paris Agreement is based on voluntary efforts to reduce greenhouse gas emissions as a key way to stem climate change. There has been some progress to reduce greenhouse gas emissions, but many would like to see greater efforts and better results.

Specifically, one goal of the Paris Agreement is to keep the increase in global average temperature to well below 2°C above pre-industrial levels. In other words, this is the amount of climate change that is believed to be endurable, rather than catastrophic. Because of the dangers of even this amount of global temperature increase, efforts are often focused on actually limiting the temperature rise to 1.5°C.

Greenhouse gas emissions are a primary reason for the existing increase in the global average temperature and the resulting climate change, so those emissions are the obvious thing to address to stem climate change. We need to reduce greenhouse gas emissions to at least the point where there is a balance between emissions and sinks (i.e. factors that absorb greenhouse gases), and we need to do this as soon as possible. To illustrate the situation, consider the concentration of carbon dioxide (a potent greenhouse gas) in the atmosphere. Carbon dioxide levels in the atmosphere have been increasing by about 2 or 3 ppm (parts per million) each year, primarily because of combustion emissions. In July 2018, carbon dioxide levels were of the order of 408 ppm at the National Oceanic and Atmospheric Administration's (NOAA) Mauna Loa Observatory in Hawaii (NASA, 2018). The clear first step is to stop this rate of annual increase in carbon dioxide; it needs to move towards zero.

The majority of greenhouse gas emissions, including carbon dioxide, are produced by developed countries. Countries that are still developing have been understandably concerned that steps to curb climate change might also curb their ability to continue their economic development. Therefore, the Paris Agreement on Climate Change was developed with due consideration for the issues of decreasing poverty and increasing health – two key aspects of development. Thus, the UN General Assembly (United Nations, 2015) adopted the resolution *Transforming Our World: The 2030 Agenda for Sustainable Development* on September 25, 2015. This resolution ties together a sustainable development agenda with the Paris Agreement. In fact, the issues of equity, sustainable development, and the effort to eradicate poverty are included in the language of the Paris Agreement (UNFCCC, 2015). The 17 Sustainable Development Goals and the goals of the Paris Agreement are both very high priorities for the United Nations and many of the member countries. Each country has different circumstances and is impacted by climate change in different ways. The Paris Agreement is formulated to allow each nation to develop its own plan to reduce greenhouse gas emissions.

The Paris Agreement also covers many other related topics, each of which relates to causes, consequences, or concomitant issues related to climate change and the overall goal of limiting global temperature increase to under 2°C. Article 7 of the Paris Agreement includes the goals of enhancing adaptive capacity, strengthening resilience, and reducing vulnerability to climate change (UNFCCC, 2015). Article 8 includes the development and use of early warning systems, emergency preparedness, and risk management to help address the impacts of climate change weather events. Article 9 addresses financial aspects of the Paris Agreement: The costs associated with both climate change effects and ameliorating those effects as per the Paris Agreement are already large, difficult to manage, and still growing. Article 10 includes the significance of developing new technology to help accomplish the goals of the Paris Agreement, and this is a topic on which there has been some success and great potential moving forward. The developments in wind and solar generation of electricity have already had significant impacts on reductions in greenhouse gas emissions. For instance, US greenhouse gas emissions in 2017 decreased 0.66% compared to the previous year (Houser and Marsters, 2018), bucking the previous pattern of year-on-year increases. This decrease was because of changes in the generation of electricity away from coal and towards wind and solar.

2.3 Impacts of Climate Change on Health and Welfare

A 2°C increase in global average temperature can sometimes sound like not an awful lot, but the range and size of its impacts are amazingly large. Of course, air temperatures are higher, but there is also more water – more polar ice melts rather than stays frozen and there is more moisture in the air (humidity). But on top of that, the temperature increase alters the weather patterns around the globe. Changed weather patterns lead to massive rainfall and flooding in some areas, drought and massive forest fires in other areas, and crop failures in yet other areas. No one is saying that climate change alone is creating all these issues – floods, droughts, fires, storms, and other weather patterns are always waxing and waning – but global climate change gooses all of these events into being stronger and more extreme. All the while, the melting ice caps raise the sea levels, chronically flooding populated areas that were once relatively safe from that risk. All together, these floods, fires, droughts, and other changes have massive human population displacement consequences, as people migrate to hopefully safer regions. These effects are often intertwined, but the following paragraphs review each of these impacts as separately as possible.

An increase in the global average temperature has the most direct negative impacts on people in places around the world where the temperature is already above the typical temperatures for human comfort. Places such as deserts, which are already a challenging human environment, have the potential for local outdoor temperatures that exceed human tolerance levels and start killing through heatstroke. Recall also that climate change will tend to increase humidity, which interacts with temperature to create even greater risks (Coffel et al., 2018). There are already areas in the world where summer heat can be deadly, and even higher temperatures and higher humidity due to climate change will impact the health of humans, pets, livestock, and other living creatures in those places.

Sea level rise is reducing the amount of land that is available for productive use. Flooding is also becoming a greater problem due to more substantial rainfall events because of climate change that result in larger floods. To understand the breadth of the impacts from sea level rise and flooding events, consider just one problematic area. In and around Houston, Texas there were major rainfall events with significant flooding on Memorial Day in 2015, in April 2016, and with Hurricane Harvey in 2017 (Kimmelman, 2017). Each of these three events was expected to occur only once in every 500 years prior to climate change. After these flooding events, some people decided to move to higher ground, and others have moved because of incentives to relocate. The total estimated recovery cost for Hurricane Harvey alone is $81 billion (Kimmelman, 2017). Houston is not alone. In just the United States there have been similar situations all along the Gulf of Mexico, across Florida, and running up the East Coast.

These repeated and severe floods reduce the desirability to build on lower land sites. In Houston, there is an effort to respond to the very large rainfall associated with Hurricane Harvey by improving the infrastructure for run-off and not rebuilding on some lower elevation sites. When a large number of homes are destroyed by weather events, some people migrate to other communities where housing is available. Although the climate change-induced sea level rise is presently still small, it has already begun to show up in the prices of coastal properties in anticipation of expected future changes (Keenen et al., 2018). Keenen et al. found that the perceptions of flood risk due to climate change are impacting real estate prices in Florida, with the greatest impact happening for the lowest elevations – coastal properties that are most likely to be impacted negatively by future sea level rise.

Climate change has impacts on water supplies which affect quality of life and public health. There are many places in the world where snow in the mountains is an important source of water for the next spring and summer. When the precipitation comes as rain, this changes the dynamics and impacts water supplies. Climate change is impacting temperature and precipitation, which has effects that change water supplies in many locations. Sea level rise may have impacts on coastal water supplies because of salt water in groundwater near the coast moving inland (Abedin et al., 2019). Abedin et al. address the nexus of climate change, water scarcity, and health. They state that "malnutrition and water scarcity may be the most important health consequences of climate change" (Abedin et al., 2019). Irrigated agriculture and crops that depend on natural precipitation are impacted by water scarcity. Climate variability impacts include drought during dry periods and floods during the wet season.

The impacts of climate change on the water cycle have been reviewed because of the great importance of water for food, environmental systems, and life in urban communities (World Bank, 2016). Water is essential for economic development and for meeting the Sustainable Development Goals. Water resources management is very important when there is water scarcity. Approximately 1.6 billion people live in countries with water scarcity (World Bank, 2016). Water storage reservoirs are needed in many locations to manage water supply, and some of these also have value for flood control.

About 30% of the available fresh water is present as groundwater. Because of water scarcity, the management of groundwater resources is already a major challenge in many locations in the world, including some parts of California and Kansas. When wells are pumped until the supply is exhausted, this creates a major challenge for a community. Water crises and the failure to adapt to climate change have been identified as two of the greatest global concerns which affect social stability (World Bank, 2016).

Because of the progress in renewable energy from wind and solar sources, the desalination of water using electricity from renewable sources can provide water supplies in coastal locations. Many countries have desalination plants that serve some of their water needs.

A hugely significant impact of climate change that we are only just beginning to understand is the migration of people due to the loss of property because of climate change. The World Bank predicts that over 140 million people will migrate because of the impacts of climate change (Rigaud et al., 2018); that migration includes movement within countries and migration from one country to another. The report expects increased migration in the future because of local changes in water availability, temperature, and sea level rise, all of which will result in areas of the world becoming non-viable (e.g. too hot to live in, prone to chronic crop failure, or simply not having enough water). And, of course, some migration of people will be due to the flat-out loss of property due to adverse weather events or sea level rise that are exacerbated by climate change.

Finding good places for climate migrants to go to, as with economic or political migrants, is becoming a significant challenge. Large cities are expected to become even larger as people move there in search of employment and better living conditions. Better wages and better opportunities are the goals of many migrants. Violence in some communities (perhaps exacerbated by the above impacts of climate change) can also contribute to migration trends.

The costs of climate change are estimated to be significantly greater if the goals of the Paris Agreement are not met (Jevrejeva et al., 2018). Trillions of dollars can be saved by reducing greenhouse gas emissions and limiting the global average temperature rise to 1.5°C. In this case (1.5°C), the sea level rise is estimated to be 52 cm by 2100 and the annual cost of this will be more than $6 trillion/year. One trillion dollars, much less $6 trillion, is difficult to understand. With a worldwide human population of about 7.5 billion, that means the cost is close to $1000 for every living person on the planet, or about $18,500 for every one of the approximately 325 million living Americans. Now consider that that is the *annual* cost of climate change by 2100, *if* we successfully limit the global average temperature rise to 1.5°C.

Byers et al. (2018) investigated several climate change risks and the management of those risks in terms of global exposure and the vulnerability of different populations. There are few regions or human populations that will not be impacted. Flooding risks predominate in coastal regions and other low-lying areas. Arid regions that struggle with water availability will tend to have more extreme problems with these same issues. The increase in overall temperature is expected to have a significant impact across a swath of regions. The greatest impacts are on the poorest populations, who will experience the impacts of climate change even at the 1.5°C level of increase in global average temperature. This helps to explain why, in Africa and South Asia, there are very large projected benefits from poverty reduction and climate mitigation.

Research to estimate future temperatures, populations, and geographical population distributions associated with climate change is being conducted (Harrington and Otto, 2018). The impacts of future temperatures on populations will depend on where they are and the local temperatures that they experience. Some amount of adaptation will clearly be necessary to enable populations to reduce risks and the impact of temperature on individual health, welfare, and crop production.

2.4 Importance of Science and Technology

One of the most promising ways to reduce greenhouse gas emissions is to develop new technologies that replace the use of fossil fuels for energy. To be feasible, these need to be cost competitive with combustion technologies. Two other very important developments that are receiving significant attention are the electrification of transportation and the replacement of the combustion of coal or natural gas to generate electricity with wind and solar energy for electric power production.

As of September 2018, the world is about to reach 4 million electric vehicles in service, when counting both automobiles and buses. With additional developments in technology, the point will soon be reached where many people will want to have an electric vehicle because it is the best alternative based on cost, quality, performance, and dependability. Electric vehicles by themselves will only help with part of the issue, though, as the generation of the electric power for those vehicles should ideally not come from fossil fuels.

Energy technology innovation in Canada has been reviewed by Jordaan et al. (2017). These authors point out that innovation helps in meeting Canada's greenhouse gas emission goals. Solar and wind energy production is similarly becoming economically as appealing as the use of fossil fuels. A continuing issue with wind and solar power, though, is the storage of energy during peak production periods (e.g. in the daytime for solar power) for later use (e.g. at night). Steady progress in energy storage technologies is being made, but the increasing number of electric vehicles can actually have a positive synergistic relationship on this front as well, with the storage of energy in vehicle batteries.

There has been great progress in the development of batteries for electric vehicles and for storage of electricity in those vehicles as part of an electrical grid. Additional developments in battery science and technology are anticipated, and further reductions in costs will lead to low-cost electric vehicles in developing countries that are powered by low-cost renewable power from wind and solar energy (Erickson, 2017). The efforts to advance energy innovation have many dimensions, some of which can have positive interaction effects, and aligning policy and technology so that they all work together effectively will be beneficial.

2.5 Progress Update

Part of the follow-through for the Paris Agreement on Climate Change is to regularly monitor its progress. The 24th session of the Conference of the Parties (COP24) to the UNFCCC was held in Katowice, Poland, December 2–15, 2018, with the goal of developing implementation details for the Paris Agreement on Climate Change (Allan et al., 2018; C2ES, 2018; Waskow et al., 2018). The conference (COP24) included developing and maintaining plans for regular communication, reporting of progress, reviewing of nationally determined contributions and recent climate impacts, stocktaking on emissions and financial actions, and aligning investments. Each country is to report its plans for nationally determined contributions by 2020 and every two years after that, following a common set of guidelines. There is an expert committee to provide assistance to countries that need help with the communication goal of each country, providing the information necessary for clarity, transparency, and understanding. There is to be voluntary cooperation on market-based approaches. Collective progress to reduce greenhouse gas emissions will be reviewed every five years. All of this tracking and promotion of progress is, of course, ultimately focused on the goals of reducing greenhouse gas emissions and keeping the global temperature increase to 1.5°C.

2.5.1 Greenhouse Gas Emissions

The Paris Agreement on Climate Change includes the goal of achieving a balance between greenhouse gas emissions and sinks such that the concentrations of these gases are no longer increasing in the atmosphere. The increase in carbon dioxide concentration shown in Table 2.1 is more uniform, whereas methane concentrations show more variation than carbon dioxide. The values of the two concentrations were taken from graphs and are approximate values. They show that there has been a significant increase with time during the past 33 years.

Nitrous oxide concentration was 0.33 ppm in 2017, which is 122% of pre-industrial values (i.e. before 1750). Atmospheric methane reached a new record high of 1.86 ppm in 2017, which is 257% of pre-industrial values. The value of 405 ppm for carbon dioxide is 146% of pre-industrial values. Most of the contribution to radiative forcing is due to carbon dioxide; however, methane contributes 17% and nitrous oxide contributes 6% (WMO, 2018). Finding ways to reduce methane emissions should also be a high priority because methane has a much greater impact than carbon dioxide on radiative forcing and climate change per mole of gas.

Since there are processes that remove greenhouse gas emissions from the atmosphere, it is not necessary to end all emissions to achieve the goal of a steady state concentration. Additionally, there are a number of components under the general term of "greenhouse gases", but one of these – carbon dioxide – is the one most often and most intensively focused on. Carbon dioxide is a significant proportion of the emitted greenhouse gases, and it has a long lifetime in the atmosphere, meriting the greater amount of attention it receives. An obvious solution to rising carbon dioxide emissions, along with greenhouse gas emissions in general, is simply to produce fewer emissions, but there are additional possibilities as well. Carbon capture and storage to reduce emissions is an additional option to address the carbon dioxide management issue. Another option is the possibility of developing ways to increase the removal of carbon dioxide from the atmosphere.

Table 2.2 shows average annual carbon dioxide emissions for 2014 in kg/person for selected countries. There are significant differences in values because developed countries such as Germany and the United States have much larger emissions compared to developing countries such as Kenya and Nigeria. Countries that have weather-related heating and air conditioning emissions have larger values compared to locations where there are lower emissions for thermal comfort. Saudi Arabia is an example of a country where air conditioning is important for thermal comfort. The United States and Australia are examples of large countries where emissions may be greater because of travel over greater distances.

TABLE 2.1

Global annual average temperature increase relative to 1880, annual average carbon dioxide concentration at NOAA's Mauna Loa Observatory, and global annual average methane concentrations.

Year	Temperature Increase (°C)	Carbon Dioxide Concentration (ppm)	Methane Concentration (ppm)
1985	0.31	346	1.65
1990	0.63	355	1.71
1995	0.64	362	1.75
2000	0.59	370	1.77
2005	0.86	378	1.77
2010	0.89	387	1.80
2012	0.81	392	1.81
2014	0.92	397	1.82
2015	1.06	400	1.83
2016	1.18	403	1.84
2017	1.10	405	1.85

Sources: ESRL, 2019; Lindsey, 2018; NASA, 2019.

TABLE 2.2

Average annual carbon dioxide emissions in 2014 for selected countries.

Country	Amount (kg/person)
Australia	15,400
Brazil	2600
Canada	15,100
China	7500
France	4600
Germany	8900
India	1700
Israel	7900
Japan	9500
Kenya	300
Netherlands	9900
Nigeria	500
Norway	9300
Philippines	1100
Russian Federation	11,900
Saudi Arabia	19,500
Sweden	4500
United Kingdom	6500
United States	16,500

Sources: The World Bank and Carbon Dioxide Information Analysis Center, Environmental Sciences Division, Oak Ridge National Laboratory, Tennessee; data.worldbank.org/ (2019).

The goals of reducing greenhouse gas emissions to balance sources and sinks and limiting the temperature rise to 1.5°C are different, but they are related (Betts and McNeall, 2018; Tanaka and O'Neill, 2018). For simple approximation purposes, we can estimate the concentration of carbon dioxide in the atmosphere that corresponds to the 1.5°C rise in temperature: 439–445 ppm. This projection holds up well across several models (Betts and McNeall, 2018).

It is possible to achieve a balance of emissions and sinks so that carbon dioxide and other greenhouse gas levels remain stable, and the global temperature rise does not exceed 2°C. This would avert the vastly greater, more expensive, and more deadly impacts of climate change, compared to present conditions. Reducing greenhouse gas emissions is very important, and the number of people in the world who understand this and are working to accomplish the goals of the Paris Agreement is increasing. There are scientific and technological advances that can be implemented now to help the transition to renewable energy for generating electricity and transportation. More research, development, and advances are on the horizon if we continue to place appropriate attention on this issue.

2.5.2 The Goal of 1.5°C

The goal of keeping the rise in global average temperature below 1.5°C relative to pre-industrial temperatures is very important to the quality of life for billions of residents on this planet. The global average temperature has been increasing every decade since 1970 (Lindsey and Dahlman, 2018; NASA, 2019). Furthermore, there has not been a uniform increase with time for the annual average global temperature; it has been accelerating. Relative to 1880, the annual average global temperature has increased, as shown in Table 2.1. According to the Intergovernmental Panel on Climate Change, the global average temperature has already increased by about 1°C, and it is expected to hit the 1.5°C mark between 2030 and 2052, given the current rate of change (Masson-Delmotte et al., 2018). In recent years, the increase in global average temperature has been about 0.2°C per decade. What all this means is that significant action must be taken now to reduce greenhouse gas emissions so that the rise in global average temperature slows down immediately and has a chance to stay below 1.5°C.

References

Abedin, M.A., Collins, A.E., Habiba, U., and Shaw, R. (2019). Climate change, water scarcity, and health adaptation in southwestern Bangladesh. *International Journal of Disaster Risk Science* 10: 28–42.

Allan, J., Antonich, B., Bansard, J. et al. (2018). Summary of the Katowice Climate Change Conference: 2–15 December 2018. *Earth Negotiations Bulletin*, 12: 747, International Institute for Sustainable Development, 18 December, 2018; enb.iisd.org/

Betts, R.A. and McNeall, D. (2018). How much carbon dioxide at 1.5°C and 2°C? *Nature Climate Change* 8: 546–553.

Byers, E., Gidden, M., Leclere, D. et al. (2018). Global exposure and vulnerability to multi-sector development and climate hotspots. *Environmental Research Letters* 13: 1–14; doi.org/10.1088/1748–9326/aabf45.

Coffel, E.D., Horton, R.M., Sherbinin, A.D. (2018). Temperature and humidity based projections of a rapid rise in global heat stress exposure during the 21st century. *Environmental Research Letters* 13: 1–9; doi.org/10–1088/1748–9326/aaa00e.

C2ES (2018). *Outcomes of the U.N. Climate Change Conference in Katowice*. Center for Climate and Energy Solutions; C2ES.ORG/

Erickson, L.E. (2017). Reducing greenhouse gas emissions and improving air quality: Two global challenges. *Environmental Progress and Sustainable Energy* 36: 982–988.

ESRL (2019, May 5). *Trends in Atmospheric Methane*. NOAA Earth System Research Laboratory; esrl.noaa.gov/

Harrington, L.J. and Otto, F.E.L. (2018). Changing population dynamics and uneven temperature emergence combine to exacerbate regional exposure to heat extremes under 1.5°C and 2°C of warming. *Environmental Research Letters* 13: 1–10; doi.org/10.1088/1748–9326/aaaa99.

Houser, T. and Marsters, P. (2018, March 29). *Final US emission numbers for 2017*. Rhodium Group, New York; https://rhg.com/

Jevrejeva, S., Jackson, L.P., Grinsted, A. et al. (2018). Flood damage costs under sea level rise with warming of 1.5°C and 2°C. *Environmental Research Letters* 13: 1–11; doi.org/10.1088/1748–9326/aacc76.

Jordaan, S.M., Romo-Rabago, E., McLeary, R. et al. (2017). The role of energy technology innovation in reducing greenhouse gas emissions: A case study of Canada. *Renewable and Sustainable Energy Reviews* 78: 1397–1409.

Keenen, J.M., Hill, T., and Gumber, A. (2018). Climate gentrification: From theory to empiricism in Miami-Dade County, Florida. *Environmental Research Letters* 13: 1–11; doi.org/10.1088/1748–9326/aabb32.

Kimmelman, M. (2017, November 11). Lessons from Hurricane Harvey: Houston's struggle is America's Tale. *New York Times*.

Lindsey, R. (2018, August 1). Climate Change: Atmospheric Carbon Dioxide. *NOAA Climate.gov*; www.climate.gov/

Lindsey, R. and Dahlman, L.A. (2018, August 1). Climate Change: Global Temperature. *NOAA Climate.gov*; www.climate.gov/

Masson-Delmotte, V., Zhai, P., Portner, H.O. et al. (2018). Summary for Policy Makers. *Global warming of 1.5 C. World Meteorological Organization*. Geneva, Switzerland; wmo.int/

NASA (2019, May 3). *Global Temperature*. NASA Global Climate Change; nasa.gov/

NASA (2018). *Vital Signs of the Planet*. NASA Global Climate Change; https://climate.nasa.gov/

Rigaud, K.K., Sherbinin, A.D., Jones, B. et al. (2018). *Groundswell: Preparing for Internal Climate Migration*. World Bank, Washington, DC: www.world bank.org/

Tanaka, K. and O'Neill, B.C. (2018). The Paris agreement zero-emissions goal is not always consistent with the 1.5°C and 2°C temperature targets. *Nature Climate Change* 8: 319–324.

United Nations (2015). *Transforming Our World: The 2030 Agenda for Sustainable Development*. United Nations, New York.

UNFCCC (2015). *Paris Agreement. United Nations Framework Convention on Climate Change*. United Nations.

Waskow, D., Dagnet, Y., Northrop, E., and Thwaites, J. (2018, December 21). COP24 Climate Change Package Brings Paris Agreement to Life. *World Resources Institute*; www.wri.org/

WMO (2018, November 20). Greenhouse gas levels in atmosphere reach new record. *World Meteorological Organization Press Release 22112018*; public.wmo.int/

World Bank (2019). *Carbon Dioxide Emissions*. The World Bank; data.worldbank.org/

World Bank (2016). *High and Dry: Climate Change, Water, and the Economy*. World Bank, Washington, DC; www.worldbank.org/

3

Urban Air Quality

With contributions from Ronaldo G. Maghirang

Abstract

Air quality is a major environmental issue, with ambient air quality in large cities and indoor air quality (where solid fuels are a source of energy) being two aspects of air quality that have major significance. The social cost of illness and early mortality from poor air quality exceeds $4 trillion per year. A particular risk is particulates in the air smaller than 2.5 micrometers (PM2.5) that impact health by entering the lungs and finding their way into locations where they have health effects. Nitrogen oxides, ozone, sulfur dioxide, and volatile organic compounds in ambient air also have health impacts. Efforts to improve air quality by reducing the amount of electricity generated from the combustion of fossil fuels and by electrifying transportation are in progress in many parts of the world. Household air pollution can be reduced by cooking and heating with gas or electricity rather than solid fuels. Improving air quality is a high priority in many locations and there is a need to continue to take significant and decisive actions on this front. Reducing greenhouse gas emissions by 80% by 2050 will be beneficial to communities that are working to improve air quality.

3.1 Introduction

Air quality in cities has impacted human health and quality of life for about as long as cities have existed, along with the combustion processes used for cooking, heating, and industry. Air pollution in cities is a global concern, with welfare costs that exceed $4 trillion per year (Erickson et al., 2017; UNICEF, 2016). Two important sources of air pollution are emissions from transportation and open burning. Combustion of solid fuels used for cooking contributes significantly to household air pollution and urban air quality. Combustion of coal for industrial production and power generation contributes to air pollution as well.

The total number of deaths attributed to air pollution is more than 6 million/year (Baklanov et al., 2016; HEI, 2018). Ambient air pollution contributes to more than 3 million deaths per year, and household air pollution associated with combustion for cooking and heating leads to more than 3 million deaths per year (WHO, 2018). Modern urban cities are home for more than half of the inhabitants of the world, and megacities with 10 million or more residents account for about 10% of the global population (Baklanov et al., 2016). The air quality is poor in many of these large cities.

3.2 Air Pollutants

Air pollution is a general term, and within that broad category there are several distinctions that can be important. One distinction is the size of the particles making up the pollution; another distinction is the composition of the air pollution; yet another is the location of the pollution. Each of these distinctions carries with it a slightly different set of implications and also different prescriptions for how to deal with that form of pollution. In terms of size, particulates in air that are smaller than 2.5 micrometers in equivalent aerodynamic diameter (PM2.5) are some of the more important air pollutants because they are strongly associated with health effects. In terms of content, nitrogen dioxide (NO_2), carbon monoxide (CO), and sulfur dioxide (SO_2) are present as air pollutants in urban air because of combustion processes. Other types of pollution, such as smog and ozone (O_3) are formed due to interactions between these harmful substances and other elements in the air. For example, O_3 is formed in the air through photochemical processes because of nitrogen oxides and volatile organic compounds (VOCs) that are present with sunlight in the air. These pollutants are also important to health and are included in the calculation of the air quality index (AQI) that has been used to communicate air pollution health risks to the public (Erickson et al., 2017; USEPA, 2018).

The AQI translates air pollutant concentration data into numbers and colors that help individuals to know if they need to modify their activities. The AQI has six color-coded categories: Good (green), moderate (yellow), unhealthy for sensitive groups (orange), unhealthy (red), very unhealthy (purple), and hazardous (maroon). Since particulate matter, nitrogen dioxide, and ozone are most important for urban air quality and health, values for these three pollutants are shown in Tables 3.1 and 3.2. Values for the National Ambient Air Quality Standards (NAAQS) are shown in Table 3.1, while the range for each AQI category is shown for each pollutant in Table 3.2. The primary standards in Table 3.1 are meant

TABLE 3.1

National Ambient Air Quality Standards in the USA for particulate matter (PM2.5, PM10), nitrogen dioxide (NO_2), and ozone (O_3).

Pollutant		Primary/ Secondary	Averaging Time	Level	Form
Nitrogen Dioxide (NO_2)		primary	1 hour	100 ppb	98th percentile of 1-hour daily maximum concentrations, averaged over 3 years
		primary and secondary	1 year	53 ppb	Annual mean
Ozone (O_3)		primary and secondary	8 hours	0.070 ppm	Annual fourth-highest daily maximum 8-hour concentration, averaged over 3 years
Particulate Matter	$PM_{2.5}$	primary	1 year	12.0 μg/m^3	Annual mean, averaged over 3 years
		primary and secondary	24 hours	35 μg/m^3	98th percentile, averaged over 3 years
	PM_{10}	primary and secondary	24 hours	150 μg/m^3	Not to be exceeded more than once per year on average over 3 years

Sources: Erickson et al., 2017 and USEPA, 2018.

to protect public health, while the secondary standards protect public welfare. Damage to plants and animals is also included as part of the secondary standards (USEPA, 2018).

The values in Tables 3.1 and 3.2 are for the United States. The WHO has a guideline value of 10 micrograms per cubic meter (μg/m^3) for the annual average value of PM2.5. The values of PM2.5 in the air vary with time of day and location in a city; the WHO value of 10 μg/m^3 is an annual mean value (WHO, 2016). There are also air quality standards for the countries of the European Union (EC, 2017) and for many other countries.

3.2.1 Particulate Matter in Urban Air

Particulate matter in urban air has the greatest health impact among all the pollutants. The US NAAQS for PM2.5 is 12 μg/m^3 for the annual average value. The AQI value is 50 for PM2.5 = 12, which is the transition value from green (good) to yellow (moderate) for the AQI. When the AQI = 100, the value of PM2.5 is 35.4, which corresponds to the 24-hour average value shown in Table 3.1.

The particulate matter in large cities has become a global problem of great importance. The number of megacities with more than 10 million people has been increasing, and the number of vehicles in those cities has been

TABLE 3.2
Air quality index (AQI).

AQI Categories	Health Concern	Ozone (ppm)	PM2.5 (μg/m³)	NO_2 (ppb)
Good (up to 50)	Air quality is satisfactory.	0–0.054	0–12.0	0–53
Moderate (51–100)	Air quality is acceptable; however, unusually sensitive people should consider reducing prolonged or heavy exertion.	0.055–0.070	12.1–35.4	54–100
Unhealthy for Sensitive Groups (101–150)	People with heart and lung disease, older adults and children are at a greater risk from exposure to ozone and particulate matter.	0.071–0.085	35.5–55.4	101–360
Unhealthy (151–200)	Everyone may begin to experience some adverse health effects, and those in the sensitive groups may experience more serious health effects.	0.086–0.105	55.5–150.4	361–649
Very Unhealthy (201–300)	Everyone may experience more serious health effects.	0.106–0.200	150.5–250.4	650–1249
Hazardous (301–500)	Entire population is more likely to be affected.	–	250.5–500.4	1250–2049

Sources: Erickson et al., 2017 and USEPA, 2018.

correspondingly increasing. Most megacities have annual average values of PM2.5 that exceed the WHO guideline value of 10 μg/m³ – for example, values of 55 for Beijing, 66 for Cairo, 13 for Chicago, 170 for Delhi, and 19 for Los Angeles have been reported (Erickson et al., 2017; Erickson and Jennings, 2017; Krzyzanowski et al., 2014). Tokyo is one of the largest megacities, with a population of about 38 million in 2016 (UN, 2016) and an average annual PM2.5 value of about 14 μg/m³ (Cheng et al., 2016). According to Cheng et al. (2016), the largest values of PM2.5 are in Delhi (143), Cairo (109.6), Xi'an (102.2), Tianjin (95.7), and Chengdu (89.4). The smallest values are in Miami (6.7), Toronto (8.4), New York (9.1), Madrid (9.8), and Philadelphia (10.3). These are all average annual values in micrograms per cubic meter for cities with a population of 5 million or more. For 338 big cities in China with PM2.5 data, annual average values range from 11–125 μg/m³ in 2015 and from 12–158 μg/m³ in 2016; the average value for all of these cities was 50 μg/m³ in 2015 and 47 μg/m³ in 2016 (Lin et al., 2018). In China, the combustion of coal is the source of 46% of the PM2.5 in summer and 35% in the winter (Ma et al., 2017).

The composition of the PM2.5 has been investigated and reported: About 30% is organic matter, 7% is elemental carbon, 36% is SNA which includes sulfate, nitrate, and ammonium compounds, 10% is soil, and 19% is unidentified (Cheng et al., 2016). Sources of organic matter in this air are from primary combustion emissions of gasoline and diesel vehicles and other combustion sources. Elemental carbon concentrations in PM2.5 are greater for diesel emissions compared to gasoline emissions. Secondary organic aerosols are also important. Smoke from the combustion of wood can be important in cities where wood is used for winter heating (Cheng et al., 2016). In three European cities – Amsterdam, Erfurt, and Helsinki – local combustion processes and secondary particle formation were the most important source categories of PM2.5 (Vallius et al., 2005).

Although average annual values provide a good overall assessment, the concentrations of PM2.5 vary with position and time. A good rain washes particulates from the air and improves air quality. Particulates are also moved by wind, and air quality is impacted by the typical airflow associated with the location. Los Angeles is an example of a city where the mountains and hills reduce airflow and therefore tend to reduce air quality (Parrish et al., 2016). The importance of wind velocity is also evident in the results for PM2.5 in Gansu Province in China. In Gansu Province, there are significant emissions from industrial plants and coal-burning power plants, and variations with time of day are also significant in these results (Filonchyk et al., 2016).

Lastly, the size distribution and composition of the particulates are important beyond just if they are larger or smaller than PM2.5. Vehicle emissions from engines include many particulates that are smaller than 700 nanometers. These very small particulates find their way deep into people's lungs, allowing toxic compounds to pass into the bloodstream and find their way to the brain (Erickson et al., 2017; Block et al., 2012).

3.2.2 Nitrogen Oxides from Combustion

There are three nitrogen oxides – nitrous oxide (N_2O), nitric oxide (NO), and nitrogen dioxide (NO_2) – that are found in air as gases. Nitrogen dioxide is the most important of these air pollutants because of concentration and health impacts. Since more than one nitrogen oxide is often present, "NO_x" is used to refer to mixtures of these different nitrogen oxides.

Tables 3.1 and 3.2 show the units for nitrogen dioxide in parts per billion (ppb) in the air. This is a volume fraction or mole fraction unit. The gas concentration can also be in mass/volume units. For nitrogen dioxide, 100 ppb is equal to about 188 $\mu g/m^3$ at a pressure of 1 atmosphere and a temperature of 25°C.

Nitrogen oxides are formed during combustion processes at high temperatures, with about 47% being from transportation and 34% from power production and energy use in industry (EEA, 2018). Nitrogen dioxide gas is an irritant which impacts health. Additionally, nitrogen oxides are involved in the formation of smog and ozone. Nitrogen dioxide combines with water to form acid rain, which has environmental impacts that include damage to vegetation. Nitrogen oxides can react with ammonia and other air pollutants to form solids such as ammonium nitrate and thus become particulate matter in the air (EEA, 2018; Wang et al., 2015). Nitrogen in particulate matter in Beijing, China was mostly attributed to coal combustion, vehicle emissions, dust, and animal waste (Wang et al., 2017).

Ammonia gas is an important pollutant in urban air in some locations because it contributes significantly to secondary particle formation when it reacts with oxides of nitrogen or sulfur to form small solid particles. The PM2.5 concentration in Shanghai includes sulfate-nitrate-ammonium aerosols because ammonia reacts with oxides of sulfur to form ammonium sulfate and oxides of nitrogen to form ammonium nitrate (Wang et al., 2015). Shanghai has a population of about 24 million, and haze pollution is present in the cold of winter and in spring due to particulate matter that is emitted by combustion processes and from the formation of secondary particulates in the air. The ammonia emissions in Shanghai are from industrial, agricultural, and other sources. Concentrations above 10 ppb are present in some parts of Shanghai. Reductions in ammonia emissions have the potential to contribute to reduced PM2.5 in the urban air in Shanghai (Wang et al., 2015).

3.2.3 Smog and Ozone

Smog and ozone formation in polluted air is a major problem in many large cities. As mentioned previously, nitrogen oxides react with VOCs such as vapors from fuels and solvents in sunlight to produce ozone and other compounds. The level of ozone in air can be reduced by addressing the sources of nitrogen oxides and VOCs so that the concentrations of these reactants in the air are lowered. In Los Angeles, for instance, ozone levels were as high as 600 ppb in the 1960s; however, as a result of continuing efforts to reduce the concentration of sources, ozone levels were reduced to below 200 ppb by about 1998. Ozone levels in Los Angeles are continuing to decrease (Baklanov et al., 2016; Parrish et al., 2016), although the values remain above the new national ambient eight-hour ozone standard of 70 ppb during many days each year (Parrish et al., 2016).

Air pollution has been recognized as a very important issue in Los Angeles, California since the 1950s. Parrish et al. (2016) report that there has been significant progress in reducing concentrations of ozone, VOCs, NO_x, and PM2.5 in Los Angeles and the surrounding communities because of a concerted effort to improve air quality. The progress in Southern California has been accomplished because of long-term efforts to control all air pollution sources, unified policies, and consistent and effective enforcement (Parrish et al., 2016). Controlling emissions from transportation vehicles is particularly important, and California has significant incentives to encourage the electrification of transportation as part of the effort to reduce greenhouse gas emissions and improve air quality (CARB, 2017).

3.3 Household Air Pollution

Energy production and transportation based on fossil fuels has long been recognized as being a big part of the problem of air pollution production. More recently, some attention has also been directed towards air pollution within the household. The use of solid fuels for heating and cooking is one of the important sources of air pollution in many countries, particularly in developing areas. Cooking over an open fire or with a stove that is designed for solid fuels results in air pollution. That pollution, moreover, is highly localized (inside the house), so the degree of exposure can be very high for the house's residents. A recent estimate reported that about 2.45 billion people are exposed to air pollution from the use of solid fuels for cooking (HEI, 2018). The largest numbers of people cooking with solid fuels are in India and China, whereas the largest percentages of the population cooking with solid fuels are in African countries such as the Democratic Republic of the Congo and Ethiopia. The good news is that the percentage of people cooking with solid fuels is decreasing, with more rapid declines in South Asia and East Asia (HEI, 2018).

The main reason to replace solid fuels with gas or electricity for cooking is the health impacts of the air pollution from solid fuels. More than 3 million people die from household air pollution each year, and combustion gases from burning solid fuels are the main source of pollutants (Goldemberg et al., 2018; Baklanov et al., 2016). The concept of changing from cooking with solid fuel to cooking with electricity – particularly where the electricity is generated without combustion – is included as one aspect of sustainable urban development by Stewart et al. (2018). With the lower cost of solar-generated electricity, the practicality of using off-grid solar energy to power electric cooking in rural areas where the regular electricity grid is not available is becoming possible.

3.4 Impact of Urban Air Pollution on Health

As mentioned, air pollutant concentrations can reach unacceptable levels and lead to health impacts in urban areas because of the many industrial, commercial, and mobile sources there. In global terms, the burden of disease associated with urban air pollution has been increasing in direct relation to these increases

TABLE 3.3

Deaths per 100,000 people per year associated with air pollution, and median urban annual concentration of PM2.5.

Country	PM2.5 Concentration (μg/m^3)	Deaths/100,000
China	59	76
Egypt	101	51
India	66	49
Japan	13	24
Netherlands	15	24
Nigeria	38	28
Turkey	35	44
United Kingdom	12	26
United States of America	8	12

Source: WHO (2016).

in concentrations of pollutants. Population-weighted PM2.5 concentrations increased by 18% from 2010 to 2016 (HEI, 2018). In 2016, ambient PM2.5 contributed to 4.1 million deaths and to 106 million disability-adjusted life years (DALYs, an estimate of the loss of healthy life expectancy in years). Table 3.3 reports values for the median urban PM2.5 concentration and the number of deaths per 100,000 people per year for several countries (WHO, 2016).

Erickson et al. (2017) reviewed the literature on air quality and health, finding that the social costs of air pollution are of the order of $5 trillion per year. These costs include medical costs, lost work time, and premature deaths. However, because many people in the developing world die with little or no life insurance, there is loss of life in many cases without appropriate compensation for the value of the life. Furthermore, although everyone is at an increased risk of health effects from exposure to air pollutants, children, pregnant women, and older people are more vulnerable. The most frequent air quality diseases are heart disease and stroke, asthma, chronic obstructive pulmonary disease, lung cancer, and respiratory infections (Erickson et al., 2017; HEI, 2018). Particulate matter in air is the most significant of the air pollutants in terms of its effects on human health (HEI, 2018).

There is now evidence that small particulates enter the lungs, move into the blood, and reduce cognitive function in the brain. In particular, there are higher rates of Parkinson's disease, dementia, and autism where polluted urban air is breathed. There is also a strong correlation between autism in children and maternal exposure to air pollution during the third trimester of pregnancy. The air pollution associated with freeways, in particular, appears to have an impact on autism rates. PM2.5 and ozone air pollution experienced during pregnancy are more broadly associated with increases in adverse birth outcomes (Erickson et al., 2017). Multiple sclerosis has been identified as a disease that is exacerbated by PM2.5 in urban air. Nitrogen dioxide causes inflammation of the airways, coughing, wheezing, reduced lung function, and increased asthma attacks, according to the American Lung Association and the US EPA. Also, elemental carbon in PM2.5 may be a causal factor in the development of lung cancer (Erickson et al., 2017).

The impact of air pollution on the likelihood of people having strokes is significant, with 29.2% of the DALYs from stroke being due to air pollution, where both ambient and household pollution are included. For ambient air pollution, the value is 22% of DALYs in China and India, compared to 10.2% for higher-income countries. More than 80% of 102 million DALYs in 2013 were from countries that were included in low- and middle-income categories (Erickson et al., 2017; Feigin et al., 2016).

Diabetes is also impacted by air pollution, according to a report by Bowe et al. (2018) where a cohort of 1,729,108 participants was followed for a median time of 8.5 years. The results showed that the risk of diabetes increased with an increase in PM2.5 concentration, with about 8.2 million DALYs caused by diabetes and 206,105 deaths from diabetes associated with air pollution. The effects of PM2.5 on diabetes were greater in lower-income countries (Bowe et al., 2018). In Table 3 of Bowe et al.'s study (2018), values of age-standardized DALYs per 100,000 people associated with PM2.5 exposure are reported, with the largest values being for Pakistan (221.7), Indonesia (189.4), and India (165.5). The smaller values are for Japan (46.9), Russia (46.8), and Nigeria (73.6). The value for the USA is 80.7.

One can also look at the inverse of additional injuries, illnesses, and deaths associated with air pollution; other research shows that children mature and have better health when the air quality is improved (Gilliland et al., 2017). For instance, research in Southern California has reported on the impact of improved air quality on the respiratory health of children there. This report shows that, as the concentrations of nitrogen oxides, PM2.5, and reactive organic gases have been decreasing in Southern California, the respiratory health in children has been improving because of the better air quality. The results of 20 years of research show that the greatest decreases in air pollution in Southern California are in nitrogen oxides, PM2.5, and reactive organic gases, with some decrease in ozone concentration. Three cohorts of children (1992–1993, Cohort C; 1995–1996, Cohort D; and 2002–2003, Cohort E) participated in this research, and forced expiratory volume (FEV) and forced vital capacity (FVC) were measured annually or every other year and compared to expected values for children of each age. Annual questionnaires were also used to obtain additional information. The results indicate that improvements in lung function in children were related to decreases in air pollutants, especially nitrogen oxides and PM2.5. In the latest study with Cohort E, fewer children exhibited clinical deficits in the measured lung functions, FEV and FVC. In addition, symptoms of respiratory conditions in children were fewer, especially in children with asthma (Gilliland et al., 2017).

The continuing efforts to improve air quality in Southern California are having positive impacts on health. Because the efforts have included a number of actions to improve air quality, and there have been decreases in the concentration of several air quality variables, the improvements in health cannot be attributed to any one variable or any one change in regulations. Nevertheless, efforts such as these help to provide blueprints for other geographical locations to modify and adopt. The effort to reduce emissions from mobile sources appears to be one of the actions that has contributed to the positive results. Decreases in concentrations of nitrogen dioxide and PM2.5 appear to be important contributing factors in the health results.

A comparison of five emission sources (traffic, coal combustion, diesel exhaust, wood combustion, and dust) shows that each source of emission has different properties and therefore implications (Hime et al., 2018). Emissions from traffic are a major source of small particulates that are less than 100 nm. Elemental carbon (black carbon) is one of the substances in very small particulates that is associated with cancer and is a health concern (IARC, 2012). Diesel vehicles are one of the significant sources of elemental carbon (Hime et al., 2018). Particulate matter from vehicle emissions is known to be a significant health concern. Coal-fired power plants, by contrast, emit sulfur dioxide, which forms sulfate particulates in air. Particulates containing sulfur from coal plants are associated with respiratory mortality (Hime et al., 2018). Secondary inorganic particulates associated with coal combustion have been associated with cardio-respiratory health in a number of studies (see review by Hime et al., 2018).

Emissions from traffic, diesel exhaust, and coal-fired power plants have all been identified as important sources of PM2.5 that affect health. Emissions containing particulates with metal content have additionally been found to impact health in a number of studies (Hime et al., 2018), and coal may have metal content and be a source of metal oxides in power plant emissions. Wood burning for heat and cooking has the benefit of being a renewable resource, but air pollution associated with the emissions is a health issue (Hime et al., 2018).

In summary, the relative importance and implications of these different emissions depend on the size of the particulates and their content. The very small particulates from diesel engines that contain elemental carbon are known to be associated with health concerns because they find their way into the lungs and bloodstream. As a general conclusion, though, the implications of air pollution – across all the variations in content, concentration, and location – are consistently negative and severe.

3.5 Improving Urban Air Quality

There are a number of reasons to address urban air quality, and there are many ways to improve urban air quality. Some of these options are presented in Erickson et al. (2017). For reducing health impacts, the sources contributing to human exposure should be known. For example, based on published data for PM2.5, Karagulian et al. (2015) reported that, globally, 25% of urban air pollution is contributed by traffic, 15% by industrial activities, 20% by domestic fuel burning, 18% from natural dust and salt, and 22% from other unspecified sources. Since many pollutants are generated via combustion processes, actions to reduce combustion are important in efforts to improve air quality. These include eliminating the combustion of coal and other solid fuels as much as possible, including transitioning to electric vehicles and replacing other gasoline and diesel engines with electric power that is generated cleanly by solar, wind, hydro, or nuclear systems. When the value of better health because of improved air quality is considered, the benefits of making this transition are cost effective in many cities. The social cost of $4 trillion per year associated with poor air quality is a significant incentive to find ways to reduce combustion in urban environments. Wang et al. (2016) reviewed 15 recent publications and recommended four specific actions: 1) regulations on air emissions, 2) road traffic interventions, 3) energy generation changes, and 4) reductions in greenhouse gas emissions to reduce concentrations of PM2.5 and nitrogen oxides in the air.

One of the ways that combustion emissions can be reduced is by taking action to promote and support walking and bicycle transportation. The pathways for walking and cycling should be safe and comprehensive. It is also beneficial if these paths are away from combustion sources so the air is better. Integration of public transportation and bike paths – so that people can take their bikes on public transport, ride with other passengers, and then find a good bike path for the next part of their trip – has value. There is a need to have sufficient and convenient parking for bicycles, as well as charge stations for electric bicycles. Some communities have bicycle share programs, where individuals may make use of a bike by riding it from one bike rack to another.

Another way to help with emissions is to have more natural features and spaces. Vegetation can provide environmental services by removing pollutants from the air. Parks with plants and trees, vegetation along streets and roads, and hedges that do not reduce visibility at intersections have public air quality benefits. Abhijith et al. (2017) reviewed the literature on air pollution abatement by vegetation and reported on the benefits of several alternatives. This can fit well alongside increased support for walking and cycling; cities can develop bike paths away from the road, with vegetation to reduce noise and capture air pollutants so that citizens are encouraged to travel by bicycle.

A number of cities have low- or zero-emission zones where people walk, ride bicycles, or take electric public transportation. In some low emission zones, a fee must be paid to drive into the area. More than 200 low emission zones have been established in Europe (Erickson et al., 2017: Wang et al., 2016). One alternative to further incentivize the adoption of electric vehicles is to allow electric passenger cars and electric delivery trucks in some of these areas, as they do not alter their status as low emission zones.

Looking around the world, there are significant differences in recent trends regarding urban air quality. In many large cities in South Asia (in India, Pakistan, and Bangladesh), air quality has been declining because of increased traffic and growth of the cities. Air quality in China has received attention recently, and efforts to improve air quality have shown some progress. Part of these efforts to improve air quality in China has been a significant push to electrify transportation. In some African countries such as Nigeria, PM2.5 concentrations in urban air are increasing. In some parts of Europe and the United States, there is a continuing effort to improve air quality, which is having good results (HEI, 2018; Erickson et al., 2017). Finally, the importance of air pollution control is being increasingly recognized in India, and there are current efforts to improve the National Clean Air Programme there (NCAP, 2018).

3.6 Public Education

Few people are totally indifferent to air quality, but there is a substantial gap between the perceptions that many people have (e.g. that poor air quality is merely an annoyance) and the reality. The reality, as reviewed in this chapter, is that poor air quality is a serious, pervasive, and extensive health risk that costs trillions of dollars annually in lost lives and productivity. Urban air quality, in particular, is an issue that has the biggest impact on the largest number of people. There are some great current efforts to educate people about the importance of air quality and its impact on health, but additional public education is recommended because of how important this issue is for health and for quality of life. Public information and education is needed with respect to the current values of variables such as PM2.5, the effects of air pollution on health, the methods and actions that reduce pollutant concentrations in the atmosphere, the actions individuals can take to reduce their exposure to air pollutants, and the actions communities can take to improve air quality.

In many communities, these greater efforts to inform citizens of the current air quality conditions would be beneficial and would not need to be complicated or difficult. Making the local AQI values easily available, such as on the Internet and electronic bulletin boards, would be valuable to people overall. It would be invaluable information for those at risk of, or already dealing with, asthma and other relevant health concerns. There are now low-cost monitors that can be used to measure concentrations of PM2.5, NO_2, O_3, and other air pollutants to inform citizens of those concentration levels at busy intersections. Alongside these postings of air quality conditions, of course, there should be basic information about how to interpret the data, which would be useful as a public education lever. As citizens are better able to understand the air quality data that is presented, they can also better understand the implications of that data.

In a democratic society, citizens vote on issues to make their collective voices heard. We all provide input on which policies should (or should not) be enacted and which representatives are elected to serve us. Informed citizens are necessary for effective, participatory democracy. When communities are considering actions to take on improving air quality, there are many factors to consider: Sources and quantities of emissions, pollutant concentrations, spatial distribution, exposure to the public, and vulnerability of at-risk individuals (Cartier et al., 2015; Erickson et al., 2017). It is a complex topic, and that means that our efforts in public education need to be purposive, extensive, and ongoing.

References

Abhijith, K.V., Kumar, P., Gallagher, J. et al. (2017). Air pollution abatement performances of green infrastructure in open road and built-up street canyon environments. *Atmospheric Environment* 162: 71–86.

Baklanov, A., Molina, L.T., and Gauss, M. (2016). Megacities, air quality and climate. *Atmospheric Environment* 126: 235–249.

Block, M., Elder, A., Auten, R.L. et al., (2012). The outdoor air pollution and brain health workshop. *NeuroToxicology* 33: 972–984.

Bowe, B., Xie, Y., Li, T. et al. (2018). The 2016 global and national burden of diabetes mellitus attributable to PM2.5 air pollution. *Lancet Planet Health* 2: e301–e312.

CARB (2017). *California's Advanced Clean Cars Midterm Review*. California Air Resources Board Report; www.arb.ca.gov

Cartier, Y., Benmarhnia, T., and Brouselle, A. (2015). Tool for assessing health and equity impacts of interventions modifying air quality in urban environments. *Evaluation and Program Planning* 53: 1–9.

Cheng, Z., Luo, L., Wang, S. et al. (2016). Status and characteristics of ambient PM2.5 pollution in global megacities. *Environment International* 89–90: 212–221.

EC (2017). *Air Quality Standards*. European Commission of the European Union; http://ec.europa.eu/

EEA (2018). *European Union Emission Inventory Report 1990–2016*. United Nations Economic Commission for Europe, European Environment Agency, 2018.

Erickson, L.E. and Jennings, M. (2017). Energy, transportation, air quality, climate change, health nexus: Sustainable energy is good for our health. *Aims Public Health* 4: 47–61.

Erickson, L.E., Griswold, W., Maghirang, R.G., and Urbaszewski, B.P. (2017). Air quality, health, and community action. *Journal of Environmental Protection* 8: 1057–1074.

Feigin, V.L., Roth., G.A., Naghavi, M. et al. (2016). Global burden of stroke and risk factors in 188 countries during 1990–2013: A systematic analysis for the global burden of disease study 2013. *The Lancet Neurology* 15: 913–924.

Filonchyk, M., Yan, H., Yang, S. et al. (2016). A study of PM2.5 and PM10 concentrations in the atmosphere of large cities in Gansu Province, China, in summer period, *Journal of Earth System Science* 125: 1175–1187.

Gilliland, F., Avol, E., McConnell, R. et al. (2017). *The Effects of Policy-Driven Air Quality Improvements on Children's Respiratory Health. Health Effects Institute Report* 190, Boston, MA.

Goldemberg, J., Martinez-Gomez, J., Sagar, A. and Smith, KR. (2018). Household air pollution, health, and climate change: cleaning the air. *Environmental Research Letters* 13: 030201.

HEI (2018). *State of Global Air 2018*. Health Effects Institute, Boston, MA.

Hime, N.J., Marks, G.B., and Cowie, C.T. (2018). A Comparison of the health effects of ambient particulate matter from five emission sources. *International Journal of Environmental Research and Public Health* 15: 1206; doi:10.3390/ijerph15061206.

IARC (2012). *Diesel and Gasoline Engine Exhausts and Some Nitroarenes*. International Agency for Research on Cancer, Vol. 105, Lyson, France.

Karagulian, F., Belis, C.A., Dora, C.F.C., Pruss-Ustun, A.M. et al. (2015). Contributions to cities' ambient particulate matter (PM): A systematic review of local source contributions at global level. *Atmospheric Environment* 120: 475–483.

Krzyzanowski, M., Apte, J.S., Bonjour, S.P. et al., (2014). Air pollution in the mega-cities. *Current Environmental Health Reports* 1: 185–191.

Lin, Y., Zou, J., Yang, W., and Li, C.Q. (2018). A review of recent advances in PM2.5 in China. *International Journal of Environmental Research and Public Health* 15: 438, 1–29.

Ma, Q., Cai, S., Wang., S. et al. (2017). Impacts of coal burning on PM2.5 pollution in China. *Atmospheric Chemistry and Physics* 17: 4477–4491.

NCAP (2018). *National Clean Air Programme – India*. Ministry of Environment of India; www.moef.nic.in

Parrish, D.D., Xu, J., Croes, B., Shao, M. (2016). Air quality improvement in Los Angeles – Perspectives for developing cities. *Frontiers in Environmental Science and Engineering* 10(5): 11; doi:10.1007/s11783-016-0859-5.

Stewart, I.D., Kennedy, C.A., Facchini, A., and Mele, R. (2018). The electric city as a solution to sustainable urban development. *Journal of Urban Technology* 25: 1, 3–20.

UN (2016). *The World's Cities in 2016*. United Nations Economic and Social Affairs.

UNICEF (2016). *Clear the Air for Children: The Impact of Air Pollution on Children. United Nations International Children's Emergency Fund Report*. UNICEF, New York; www.unicef.org/

USEPA (2018). *National Ambient Air Quality Standards Table*. US Environmental Protection Agency; www.epa.gov/

Vallius, M., Janssen, N.A.H., and Heinrich, J. (2005). Sources and elemental composition of ambient PM2.5 in three European cities. *Science of the Total Environment* 337: 147–162.

Wang, L., Zhong, L., Vardoulakis, S. et al., (2016). Air quality strategies on public health equity in Europe: A systematic review. *International Journal of Environmental Research and Public Health* 13: 1196.

Wang, S.S., Nan, J., Shi, C. et al., (2015). Atmospheric ammonia and its impacts on regional air quality over the megacity of Shanghai, China. *Scientific Reports* 5: 15842; doi:10.1038/srep15842.

Wang, Y.L., Liu, X.Y., Song, W. et al. (2017). Source apportionment of nitrogen in PM2.5 based on bulk nitrogen isotope 15 signatures and a Bayesian isotope mixing model. *Tellus B* 69: 1299672.

WHO (2016). *Ambient Air Pollution: A Global Assessment of Exposure and Burden of Disease*. World Health Organization.

WHO (2018, May 8). *Household Air Pollution and Health*. World Health Organization.

4

Electrification of Transportation

Abstract

The electrification of transportation is a key development for society because of its impact on reducing greenhouse gas emissions and improving air quality. The economic appeal of electric vehicles has continuously increased with the rapid decline in battery prices over the past several years, with average prices of less than $200/kWh in 2018, and projected future prices of less than $50/kWh. With reduced battery prices, electric cars and electric bikes will become attractive options for those looking for inexpensive transportation. In 2018, the number of EVs in service (cars, buses, and trucks) increased to more than 5 million, with a large fraction of these being in China. There were more than 100 million electric bicycles in service in 2018, with more than 60% of these being in China. Electric buses have significant air quality benefits which have helped to justify their purchase in China, in California, and in major cities like London and elsewhere. Norway currently leads the world in this transition, with about 40% of new car sales being EVs in 2018. EVs that can be charged with renewable electricity also have the capability to store energy and provide benefits to grid operators and managers.

4.1 Introduction

The electrification of transportation is a major development that can significantly help in the reduction of greenhouse gases and the improvement of air quality. Its importance comes from its potential to address the two largest sources of greenhouse gas emissions: Direct emissions from transportation and potentiating shifts in the generation of electricity to cleaner methods. Although either of these sources of emissions can be tackled separately – we can shift to electric vehicles without changing the methods of electricity generation, or we can change electricity generation methods without adopting electric vehicles – the combination of these two developments is particularly powerful. This chapter reviews a number of the recent developments in electric cars, electric bicycles, electric buses, and electric trucks that have resulted in significant growth in the number of new vehicles sold annually. There are more than 100 million electric vehicles in use, as of 2018. Although the most significant market overall is plug-in electric cars, there are also a large number of electric bicycles, particularly in China. Electric buses and trucks are also rapidly developing areas with great potential, but they are not as far along in terms of implementation.

4.2 Plug-In Electric Cars

Within the last decade, battery electric vehicles (BEVs) have gone from novel curiosities to an increasingly normal option for all new car buyers to consider. As car buyers consider BEVs, they are finding that they are on a par with conventional vehicles (internal combustion engines, or ICEs) and are rapidly becoming superior in many ways. At the time of this writing, in September 2018, reported US sales of electric vehicles (EVs) in August 2018 were up more than 100% compared to August 2017 (Loveday, 2018). In Sweden, more than 9% of new car sales in August were plug-in EVs (Kane, 2018b). In the first half of 2018, 6.2% of new car sales in California were plug-in EVs, with 3.3% being all-electric and 2.9% being plug-in hybrid EVs (Pyper, 2018). Sales of hybrid vehicles without plugs have been declining for several years in California. In other words, cars such as the original Toyota Prius (which was groundbreaking in its time) are being replaced by plug-in vehicles. In August 2018, monthly sales of EVs were more than 30,000 in the USA for the first time. Global sales in June 2018 were more than 150,000, with more than one-third of these taking place in China (Loveday, 2018). Globally, more than 1 million EVs were sold in 2017 (IEA, 2018). Sales increased significantly in 2018, with about 2.1 million plug-in vehicles being delivered to customers (Irle, 2019); these include BEVs and plug-in hybrid electric vehicles (PHEVs) (cars and light trucks in the USA and Canada, and light commercial vehicles in Europe and Canada). The top five companies in terms of the number of light passenger EVs delivered in 2018 were Tesla, BYD, BAIC, BMW, and Nissan. In 2018, 93% of global EV sales were in China, Europe, and the USA (Irle, 2019).

Approximately 40 different models of EVs are available in the US market, and in August 2018, sales for eight of these models were at more than 1000 vehicles. Looking forward, there are a large number of additional new models of plug-in EVs planned across several different manufacturers. The strongest selling EV at the moment is the Tesla Model 3, which reported sales of 17,800 units in August 2018; about half of the USA's EV August sales (Loveday, 2018). Although many of the EVs are small cars, there are larger cars available and many larger models scheduled for release soon. Among the current top eight market leaders, the Tesla Model 3, Tesla S, Tesla X, Chevrolet Bolt, and Nissan Leaf are all-electric, whereas the Toyota Prius Prime and Honda Clarity Plug-in EV are PHEVs. The Chevrolet Volt is powered by electricity, but it has a gasoline motor that generates electricity to extend its range.

Many of the EVs that are sold in the USA are sold in other countries as well. And, of course, the automobile industry is global and many manufacturing operations extend across more than one country. There is significant variation in EV sales across different parts of the world, as well as across different parts of the USA. California has the strongest EV market within the USA, largely driven by the state's incentives to encourage EV ownership as a way to help improve air quality. Air quality is similarly one of the reasons for EV purchase incentives in China. Happily, these incentives, driven primarily by immediate air pollution concerns, also have the longer-term impact of helping to reduce greenhouse gas emissions more generally.

4.2.1 Advantages Associated with Electric Vehicles

It's one thing to manufacture electric vehicles, but they then need to be bought by people as their choice of transportation. As with any choice, there are advantages and disadvantages to each option. By many accounts, the current situation is somewhere around equivalence; BEVs and gas vehicles are different but are about equally appealing. The future, however, belongs to BEVs. Consider that internal combustion vehicle technology has been refined for over a hundred years, so any current advances are relatively small improvements. BEVs, though, are at an earlier stage and are rapidly improving along multiple dimensions.

So what are the major differences between a BEV and an ICE? Whereas an ICE has a number of complex engine systems (e.g. cooling system, ignition system, transmission, and exhaust system), a BEV has a large battery (or, more often now, a bank of batteries). One of the major costs of electric vehicles is the cost of the batteries. Battery costs are presently about $200/kWh, with the expectation that these prices will continue to decline and become less than $70/kWh by 2025. As battery costs decline, the EV market will become ever more attractive. Brase (2019) found that vehicle choices (electric versus gas vehicles) at the population level were strongly correlated with economic considerations.

The energy conversion efficiency is much better for EVs than for ICEs. The efficiency estimate is that 59–62% of the electrical energy supplied to an EV is converted to power at the wheels. For comparison, the efficiency estimate for a gasoline-powered car is 17–21% (EPA, 2018). This energy conversion efficiency, however, is not an immediately perceived benefit for the individual EV owner. Of more interest to a vehicle owner is their experienced cost, and the most informative estimate of that is the total cost of ownership – not only the purchase cost, but also fuel, insurance, maintenance, and other costs.

The total cost of ownership (TCO) for various vehicles has been investigated in a number of studies (Mitropoulos et al., 2017; Palmer et al., 2018). The lowest TCO when cars of different types are compared (ICEs, hybrid vehicles (HEVs), PHEVs and BEVs) turns out to vary across geographical location and the specific vehicles that are compared. The BEV had the lowest TCO in

California and the United Kingdom, whereas the ICE had the lowest TCO in Texas and Japan (Palmer et al., 2018). Again, though, this is based on specific models: The Nissan Leaf, Toyota Prius, Toyota Plug-in Hybrid Prius, Toyota Corolla, and Ford Focus Diesel (UK only) were compared. Looking at some specific locations helps us to understand this situation. In places where gasoline costs are lower (e.g. Texas), the PHEV had the lowest energy cost. However, in the United Kingdom, where petroleum costs are higher, the BEV had the lowest energy cost. PHEVs also did well in California, where there are incentives to purchase electric vehicles.

In this same comparison of vehicles, the maintenance costs were lowest for the BEVs across all cases (Palmer et al., 2018). The reason for this is very straightforward: A BEV has drastically fewer parts in its drive train. That means fewer things to maintain and fewer things that can break, that need replacement, or that can require adjustment. The maintenance costs of BEVs are actually so low that there has been some concern about the second-order effects for dealerships, repair shops, and auto parts retailers. As EVs become a larger share of the car population, these parts of the automotive infrastructure may contract.

Another, more extensive type of comparison is called a life cycle analysis, in which factors such as greenhouse gas emissions and emissions that impact air quality are included. A life cycle analysis has been used to compare values of TCO for an ICE, a HEV, and a BEV (Mitrospoulos, et al., 2017). In this study, the HEV had the lowest TCO. The environmental costs that were included were less than 10% of the TCO.

It is safe to say that these types of comparisons will continue to shift in favor of BEVs in the future. The cost of batteries is decreasing, making BEVs more competitive. At the same time, the price of petroleum, which varies significantly across locations, is generally increasing over time because the supply is finite. The price of electricity has been nearly constant during the last ten years in many locations, and prices have been falling for electricity generated by wind and solar energy. Thus, over the next 20 years, there is a good likelihood that the TCO for BEVs will decrease when adjusted for inflation.

An additional benefit of EVs is their ability to convert potential and kinetic energy into electrical energy by generating electricity during braking (i.e. regenerative braking). The potential energy of a vehicle at higher elevations can be converted to electrical energy as it comes down a mountain or hill. In regenerative braking, the motor acts as a generator and produces electricity, and that generated energy is stored in the battery, where it can be used later to power the EV. The efficiency of regenerative braking has been estimated to be about 22% in Rotterdam driving (Van Sterkenburg et al., 2011). In mountain driving, it is possible to add several kWh of energy to the battery going down a mountain in a plug-in hybrid Toyota Prius. The conversion efficiency of potential energy to electrical energy appears to be significantly more than 22% when this is done at a slow speed (Erickson, 2018). When one reviews the mountain driving experiences of others as posted on the Internet, one finds that approximately 50% of the potential energy change can be converted to electrical energy through regenerative braking when coming down a mountain at modest speeds.

An initial concern of many people who consider buying an EV is the issue of charging it. There are gas stations on corners of every town and city, but where can you charge an EV? The answer to this turns out to be nearly as simple as the question of where you charge your phone: You plug it into the electrical grid when you are at home. One of the advantages of EVs is the simplicity of connecting the car to the grid to charge the battery. EVs can recharge their batteries from a basic 120-volt outlet. For most people (with daily driving of about 40 miles), overnight charging on a standard wall outlet is sufficient. For those who have longer commutes, a garage with electric vehicle supply equipment (EVSE) charging at 240 volts is also easy and convenient. When on a road trip where one needs to stop to charge the battery, more time is needed, and so it makes sense to charge the battery while at a destination or when one stops to eat or sleep. With charging at home, an EV essentially leaves the house each day with the advantage of a "full tank". This in turn can help to alleviate "range anxiety": The fear of running out of electrical charge in an EV. Range anxiety remains a key concern for a number of people, but it is progressively becoming less of a concern as the technological innovations of EVs have increased ranges from less than 100 miles to a number of new models having ranges of well over 200 miles for a full charge.

Some recent research (Brase, 2019) indicates that important factors which predict greater intentions to buy an electric vehicle are confidence about the performance and range capabilities of EVs, along with the perceived prevalence of EVs in general, and an electric vehicle being consistent with the owner's social values. The technology improvements for EVs are already doing a good job of addressing the first set of these concerns, and the later concerns should be addressed with the larger number and variety of EV models coming out.

4.2.2 International Developments

The advantages of electric vehicles transcend all national borders, for the most part. Certain countries, however, have sped up the adoption of EVs by providing a variety of incentives to car owners. Norway is one of the global leaders with respect to incentives to encourage the purchase of EVs. Norway has a population of less than 6 million, meaning that it is not a large country compared to some others, but 96% of Norway's electricity is produced in hydroelectric power plants (Figenbaum, 2017), which means that EVs can be charged with electricity that has been generated without carbon emissions. The incentives for Norwegian EV owners include exemption from registration tax and value added tax, a reduced license fee, free use of toll roads, reduced rates on ferries, access to bus lanes, and free parking (Figenbaum, 2017). These incentives make a difference; people respond to opportunities that are beneficial to them. During the first seven months of 2018, 3738 new plug-in passenger EVs were registered in Norway; this is about 45.7% of the new car market (Kane, 2018a). For all of 2018, 40% of new car sales were EVs (Irle, 2019); no other country has a higher percentage of EVs for new car purchases. In Iceland, 17.5% of new car sales were EVs in 2018, compared to 7.2% in Sweden (Irle, 2019). A number of other locations have also implemented incentives for people to transition to electric vehicles, as well as incentives to generate electricity with renewable energy. These are often justified not only because of the need to reduce greenhouse gas emissions, but also because of their more immediate effects in improving air quality.

By country, the largest number of EVs sold in 2017 was in China, with about 580,000 new car EV sales (IEA, 2018). For 2018, the number of new light-duty EVs increased to 1.2 million in China (Irle, 2019). Worldwide, the number of light EVs in service was more than 5.4 million at the end of 2018, and the increase from the end of 2017 was more than 50%. In summary, there was good growth in EV sales in 2017, and that sales growth continued to accelerate in 2018. Globally, there are more than 50 different models of EV on sale in 2019.

4.3 Electric Bicycles

Another important recent development is the growth in the electric bicycles (e-bikes) market. E-bikes are relatively small, with batteries that are about 500 Wh in many cases, but many e-bikes allow the rider to travel at speeds that keep up with city traffic (Barnard, 2018). Although a relatively small market in the USA, e-bikes are a very large and growing part of the transportation mix in China.

4.3.1 Electric Bicycles in China

E-bike sales exceed 20 million per year in China, and they are used for inexpensive and efficient daily transportation to work and other locations. The cost of owning and operating an e-bike is of the order of 1 cent per km (Salmeron-Manzano and Manzano-Agugliaro, 2018). In 2015, worldwide sales of e-bikes were more than 40 million, with most of these in Asia, about 5% in Europe, and less than 1% in the USA. Some e-bikes are designed to provide power when the person on the bike is pedaling (pedalecs). Some e-bikes only provide power up to a given speed, such as 20 or 25 kmph. The e-bike provides more exercise than riding in a car, but less exercise than riding a regular bicycle. People often get to their destination faster with an e-bike compared to a regular bike (Berntsen et al., 2017).

One of the reasons for e-bike growth in China has to do with regulations designed to improve air quality by not allowing ICE motorcycles in some of the large cities. Lin (2016) reported that electric bicycles have become popular in China because of their flexibility to travel directly to a destination (as compared to public transportation). At the same time, the electrical energy required for e-bikes is of the order of 2 kWh/100 km, compared to about 20 kWh/100 km for an electric car. An e-bike is also inexpensive compared to owning an automobile, and it enables many people to travel to work, school, or other destinations conveniently and at a reasonable speed. Lin (2016) reported that about 22% of users have paid less than 2000 CNY for their e-bike, 24% paid between 2000 and 3000 CNY, 40% paid between 3000 and 4000 CNY, and 14% paid more than 4000 CNY (7 CNY = $1). For context, the cost of an e-bike in China is about one month's salary for some users. The data on households with an e-bike in Nanjing shows that 37.72% have one car, 22.33% have one motorcycle, and 40.69% have a bicycle. About 66% use their e-bike to commute regularly to a destination, such as work.

4.3.2 Electric Bicycles in Other Countries

People in the United States can purchase an electric bicycle for less than $3000, such as the e-bike marketed by Yamaha (Toll, 2018). In general, electric bicycles can be purchased and operated without a license and insurance in the USA, but there are both federal and state regulations that define e-bikes, and these are important to understand.

One of the significant issues with e-bikes is the regulations that owners need to follow; another is what qualifies as an e-bike. The e-bikes are expected to travel on pathways designed for bicycles where they are available, and so it is important to have power restrictions. This is also true for issues such as any local requirements for a vehicle license and insurance.

In most countries, there are regulations on the power and/or speed that define e-bikes. In the United States, the definition of an e-bike is:

> a two- or three-wheeled vehicle with fully operable pedals and an electric motor of less than 750 W (1 h.p.), whose maximum speed on a paved level surface when powered solely by such a motor while ridden by an operator who weighs 170 pounds (77 kg) is less than 20 mph (32 kmph). (NCSL, 2018)

A Class 1 e-bike is a bicycle that is equipped with a motor that provides assistance only when the rider is pedaling and that ceases assistance when the speed of the bicycle is 20 mph. A Class 2 e-bike has a motor that can be used exclusively to propel the bicycle up to a speed of 20 mph. A Class 3 e-bike provides assistance only when the rider is pedaling, and only up to a speed of 28 mph (45 kmph) (Bike Law, 2017; NCSL, 2018). The manufacture and sale of e-bikes that meet the US federal definition are regulated by the Consumer Product Safety Commission and must meet bicycle safety standards (NCSL, 2018). The electric two-wheeled vehicles that have more power than those described above are treated as motor vehicles that must meet the requirements of other motor vehicles.

In the USA, there are differences from state to state with regard to the rules and regulations that must be followed in operating e-bikes (NCSL, 2018; TREC, 2018). Some states require a license to operate an e-bike, whereas others do not. Registration of e-bikes is required in some states but not in others. Even pedals are required in some states but not in others (TREC, 2018). Helmet requirements vary from state to state (across all manner of cycles) and may be required only with the Class 3 e-bikes in some states.

There are larger two-wheeled electric vehicles that do not qualify as e-bikes and are expected to drive on roads rather than bike paths. Safety is an important issue for e-bikes. Following riding regulations is important. Wearing a helmet is not required in all parts of the world, but there are good safety reasons to wear one anyway. One of the problems with e-bike riders is that they and regular bike riders do not follow all the rules for safe riding, and a number of accidents occur because of riders not following regulations (Lin, 2016)

4.4 Electric Buses

Electric buses share many of the characteristics and advantages of electric cars, but there are some additional considerations. Buses are, of course, larger than cars, but they also carry more people and typically have fixed routes of known distances. The different uses of buses in the world include transportation for many people to work, to school, and for other purposes. Distances within cities are often appropriate for the range of electric buses, and there is a transition to electric buses underway in many cities. China moved forward with the purchase of about 100,000 electric buses per year in 2016, 2017, and 2018, and there is significant progress towards the electrification of their city bus fleets. In the USA, there are many school buses that have well-defined schedules and modest distances where electric buses can be put into service.

TCO analysis shows that the purchase price is greater for electric buses than for ICE buses, but energy and maintenance costs are less, making the TCO comparable (Lajunen and Lipman, 2016; Tong et al., 2017; Rogge et al., 2018). Like electric cars, the batteries in electric buses are a major part of their cost, but as battery costs decrease, electric buses will become even more competitive. As the electrical grid is transformed to reduce the amount of carbon emissions associated with electricity generation, electric buses will have greater benefits with respect to the reduction of greenhouse gas emissions. When urban air quality is considered, electric buses are an ideal choice in large cities with poor air quality because emissions from diesel buses affect air quality and health. Where electric buses have been introduced, those who ride them have appreciated the noise reduction and the better air quality.

In Europe, there is a coordinated effort to encourage the transition to electric buses through the European Clean Bus Deployment Initiative, which was started in 2016 (Keizer et al, 2018). A declaration of intent from this initiative is to promote the deployment of clean, alternatively fueled buses, with an emphasis on zero-emission buses. Local and regional leaders are working cooperatively to reduce the emissions and noise associated with bus transportation. The collaborative action of transport authorities, energy providers, bus and grid operators, and manufacturers is advocated through constructive meetings and communications. The goal of 30% zero- or low-emission buses by 2025 in Europe is also included in this declaration. Some cities have announced plans to stop purchasing diesel buses (Copenhagen in 2014, London in 2018, Berlin and Oslo in 2020). Other cities (Athens, Madrid, and Paris) have plans to completely remove all diesel buses by 2025 (EC, 2018).

In India, there is significant interest and some progress with respect to electric buses. A comprehensive report, 'Electric Buses in India' (GGGI, 2015), indicates that there is a strong interest in electric buses because of the need to improve air quality and reduce greenhouse gas emissions. Reduction in noise and improved energy efficiency are identified as additional benefits. The report includes an assessment of the benefits and costs. India has a large population, and the transportation vehicle market has been increasing rapidly towards India being the third largest in terms of numbers of vehicles. There are cities with electric buses and electric bus manufacturing in India (Saini, 2018; Dhabhar, 2018), and there is a goal of having 100% electric buses in cities by 2030. Delhi currently has a plan to acquire 700 electric buses, and the number of electric buses in India has been growing since 2014, when the first electric buses were put into service in Bangalore (Saini, 2018).

One of the issues with respect to electric buses is when and where to charge them. In some cases, the plan is to charge them at night. In other cases, the buses are removed from service during the day and charged – for example, rotating electric buses or supplementing with ICE buses until battery capacity reaches the necessary size for continuous daily use. A developing technology that also has good potential here is the use of wireless charging while the bus stops to discharge and pick up passengers.

4.5 Electric Trucks

Progress with respect to electrification of truck transportation is behind that of auto or bike transportation, but the efforts to electrify truck transportation are happening and can easily make significant strides based on the work done with cars and buses. There is a plan to include electric trucks in the effort to reduce greenhouse gas emissions by transitioning to land transportation that has little or no carbon emissions as part of the response to the Paris Agreement on Climate Change (Ambel, 2017).

The electrification of garbage trucks and a wide variety of urban delivery trucks thus has good potential for rapid development (Earl et al., 2018; Taub, 2018). Electric trucks are especially competitive for stop-and-start applications such as garbage routes, set delivery routes, or even on-call delivery systems with known time windows (e.g. the human operator's work hours). MAN, Mercedes, Tesla, Renault, DAF, BYD, and Volvo all have plans to produce and market electric trucks. BYD is selling heavy-duty electric trucks as of 2018, and Workhorse Group has manufactured electric delivery trucks for UPS in the United States. In Europe, UPS is working with Arrival on electric delivery vehicles that have a range of about 150 miles on a single charge. Volvo, BYD, and Motive Power Systems have manufactured and delivered electric garbage trucks.

Urban air quality is a very good reason to restrict ICE trucks, along with their emissions, in cities. There is growing interest in improving air quality by having such restrictions, but that creates a secondary problem of how to get things delivered into and taken out of the city. The electrification of vehicles that provide services within cities, given that they have modest travel distances each day, is happening in many cities because the economics are particularly attractive when air quality benefits are considered.

4.6 Summary

Electric vehicles will continue to develop longer ranges and become more cost effective as a result of further technological developments in batteries and other features of EVs. The goal of an 80% reduction in greenhouse gas emissions by 2050 can be reached if transportation is done through vehicles that use electricity generated without carbon emissions. Although there are many dimensions that need to be attended to regarding the transition to electric vehicles, it is expected that sales of electric vehicles will continue to grow, new features that improve the quality of EVs will continue to expand the number of users, and increasing prevalence of EVs will further accelerate this transition.

References

Ambel, C.C. (2017). *Electric Truck's Contribution to Freight Decarbonization*. Transport and Environment; www.transportenvironment.org; see also *Roadmap to Climate-friendly Land Freight and Buses in Europe*.

Barnard, M. (2018). Your next motorcycle might just be an electric bike. *Clean Technica*.

Berntsen, S., Malnes, L., Langaker, A., Bere, E. (2017). Physical activity when riding an electric assisted bicycle. *International Journal of Behavioral Nutrition and Physical Activity* 14: 55.

Bike Law (2017, September 19). What you should know about electric bikes and the law. *Bike Law*; www.bikelaw.com/

Brase, G.L. (2019). What would it take to get you into an electric car? Consumer perceptions and decision making about electric vehicles. *The Journal of Psychology: Interdisciplinary and Applied*, 153: 214–236. doi:10.1080/00223980.2018.1511515.

Dhabhar, C. (2018). Electric urban bus with over 200 km range launched in India by Goldstone BYD. *Car and Bike*.

Earl, T., Mathieu, L., Corneliss, S. et al. (2018, May 17 and 18). *Analysis of long haul electric trucks in EU*. 8th Commercial Vehicle Workshop, Graz; www.transportenvironment.org/

EC (2018). *European Clean Bus Deployment Initiative*. European Commission; https://ec.europa.eu/transport/themes/urban/cleanbus_en

EPA (2018). *All-Electric Vehicles. Office of Energy Efficiency and Renewable Energy*. US Department of Energy and US Environmental Protection Agency; www.fueleconomy.gov/

Erickson, L.E. (2018). Personal experience with a 2013 Plug-in Hybrid Prius.

Figenbaum, E. (2017). Perspectives on Norway's supercharged electric vehicle policy. *Environmental Innovation and Societal Transitions* 25: 14–34.

GGGI (2015). *Electric Buses in India: Technology, Policy, and Benefits*. Global Green Growth Institute, Seoul, Republic of Korea.

IEA (2018). *Global Electric Vehicles Outlook*. International Energy Agency; www.iea.org/

Irle, R. (2019). Global EV Sales for 2018 – Final Results. *EV-Volumes – The Electric Vehicle World Sales Database*; www.ev-volumes/

Kane, M. (2018a, August 5). Norway sees minor decrease in sales in July. *Inside EVs*; https://insideevs.com/

Kane, M. (2018b). Plug-in electric cars capture 9.7% market share in Sweden in August. *InsideEVs*; https://insideevs.com/

Keizer, A., Engel, H., Guldemond, M. et al. (2018, June). The European electric bus market is charging ahead, but how will it develop? *Insights*; www.mckinseyenergy/isights.com/

Lajunen, A. and Lipman, T. (2016). Lifecycle cost assessment and carbon dioxide emissions of diesel, natural gas, hybrid electric, fuel cell hybrid, and electric transit buses. *Energy* 106: 329–342.

Lin, X. (2016). Future perspective of electric bicycles in sustainable mobility in China. PhD Dissertation, Cardiff University.

Loveday, S. (2018, September 6). August 2018 plug-in electric vehicles sales report card. *InsideEVs*; https://insideevs.com/

Mitropoulos, L.K., Prevedouros, P.D., and Kopelias, P. (2017). Total cost of ownership and externalities of conventional, hybrid, and electric vehicle. *Transportation Research Procedia* 24: 267–274.

NCSL (2018, September 26). *State Electric Bicycle Laws: A Legislative Primer, National Conference of States Legislatures*; www.ncsl.org/

Palmer, K., Tate, J.E., Wadud, Z, and Nelltjhorp, J. (2018). Total cost of ownership and market share for hybrid and electric vehicles in the UK, US, and Japan. *Applied Energy* 209: 108–119.

Pyper, J. (2018, August 23). EV sales grow to 6.2% in California, as hybrid sales decline. *Greentech Media*; www.greentechmedia.com/

Rogge, M., van der Hurk, E., Larson, A., and Sauer, D.W. (2018). Electric bus fleet size and mix problem with optimization of charging infrastructure. *Applied Energy* 211: 282–295.

Saini, P. (2018). Electric buses: The foundation for India's dream of a fully electric fleet. *Intelligent Transport*.

Salmeron-Manzano, E. and Manzano-Agugliaro, F. (2018). The electric bicycle: Worldwide research trends, *Energies* 11:1894.

Taub, E.A. (2018, June 21). Buses, delivery vans and garbage trucks are the electric vehicles next door, *Seattle Times*.

Toll, M. (2018). Move over motorcycles – Yamaha is now selling electric bicycles in the US. *Electrek*.

Tong, F., Hendrickson, C., Biehler, A. et al. (2017). Life cycle ownership cost and environmental externality of alternate fuel options for transit buses. *Transportation Research, Part D* 57: 287–302.

TREC (2018). *Electric Bicycle Laws by State and Province*; https://peopleforbikes.org

Van Sterkenburg, S., Rietveld, E., Rieck, F. et al. (2011). Analysis of regenerative braking efficiency: A case study of two electric vehicles operating in the Rotterdam area. *Research Gate*; www.researchgate.net/publication 252050989

5

Renewable Energy

Abstract

Transitioning to renewable energy is a key process for tackling both air quality and climate change issues, ultimately benefitting society. Progress is being made, and new electricity-generating capacity added in 2017 and 2018 included mostly renewable systems. In 2017, wind-generated electricity became the lowest-cost alternative in a number of markets, and more than 50% of newly installed capacity was solar power. More than 100 cities have electric power systems with more than 70% renewable generating capacity. There are also over 100 cities and more than 55 countries that have a goal of 100% renewable electricity at a future date. Developments in technology and competitive procurement have driven down the cost of new renewable wind generating capacity to about $0.03/kWh in many locations. The very significant progress in improving wind and solar generating technologies has been very important in reducing the cost of electricity from renewable sources. We still need to improve and expand the use of these renewable energy technologies, but there are immense benefits to be realized in our efforts to reduce greenhouse gas emissions and improve air quality.

5.1 Introduction

The previous chapter, on the electrification of transportation, was about how to address both immediate emissions from vehicles and "potentiating shifts in the generation of electricity to cleaner methods". This chapter now covers this latter part: Shifting to cleaner, renewable energy. Renewable energy sources tend, inexorably, to be cleaner than fossil fuels. Part of this is because of the nature of fossil fuels (i.e. involving a combustion process that emits large quantities of pollutants), and part of this is because renewable fuel sources usually take less work to continuously restock.

Thus, one of the more significant Sustainable Development Goals around the world is to transition to renewable energy. At the end of 2017, there were 57 countries that had 100% renewable electricity targets (Zervos, 2018). In fact, the number of cities that are powered by at least 70% renewable electricity increased from 41 to 101 in 2017. This included Auckland, Brasilia, Nairobi, and Oslo (Zervos, 2018).

Renewable energy policies and advances in technology have contributed to the excellent progress made in 2017 and 2018. Some of this progress has been achieved using technologies such as hydroelectric power generation. There are many locations where falling water is used to generate electricity or provide power for other purposes. Since most of the locations where water can easily be used to generate electricity have been developed, much of the focus on new developments with respect to renewable energy is on wind and solar energy. Accordingly, the bulk of this chapter focuses on these two forms of renewable energy.

Good progress is being made with respect to new developments in wind and solar energy. In 2017, renewables accounted for about 70% of the additions to global power-generating capacity (Sawin et al., 2018). As with hydroelectric power, generating both wind and solar power depends on natural processes (wind velocity and solar radiation). Unlike hydroelectric power, these resources are very widely available, yet they also vary across days of the year and time of day (i.e. the level of power production is not controlled by the owner). Nevertheless, because of falling costs, both of these technologies are now competitive with other energy sources in many parts of the world. Because of decreases in the costs of wind and solar energy production, the transition to renewable energy often has a positive regional economic impact. Many landowners benefit from having windmills installed on their property, and there are employment opportunities associated with installing wind farms and adding solar panels to buildings or parking lots. A recent review of regional economic impacts associated with renewable energy projects is available from Jenniches (2018).

There are a number of locations in the United States (including California, Hawaii, New York, and Washington) where there are plans to generate all electricity without carbon emissions at a future date; however, at present there is still significant use of coal in the United States and many other countries (Roselund and Weaver, 2018). One reason for the slowness of the transition to renewable energy for electricity generation is because of the long lifetimes of electricity-generating equipment. Some coal plants can be economically viable to operate for more than 40 years, and replacement plans depend on both the service life of the equipment and actual power generation economics. This means that cheaper renewable energy is a huge step, but not enough to make a full transition to 100% renewable energy. Some fossil fuel plants are retired because new technologies are able to provide power at a lower cost, but this sometimes only happens once the old plant is near the end of its useful life. An additional consideration for these decisions needs to be air quality; some plants increasingly should be retired not just due to obsolescence but in order to improve air quality.

There are both locations and companies that are looking forward to a future that is entirely based on renewable energy. The locations include cities that have set a 100% renewable target for their electricity generation (Sawin et al.,

2018). Companies that supply solar and wind energy equipment, of course, support more development of renewable energy, but many utility companies are also increasingly realizing the need to adapt to the times. For example, Xcel Energy has a goal to reach 80% zero-carbon emissions by 2030 and 100% zero-carbon by 2050 (Roselund and Weaver, 2018).

5.2 Solar Energy

Photovoltaic (PV) technology that produces electricity using solar panels has progressed rapidly and is now competitive with other generating methods in many parts of the world. Accordingly, there is significant installation of solar generating capacity in many countries. At the end of 2017, solar PV global capacity was about 402 GW, and 98 GW were added in 2017. At the end of 2018, solar PV global capacity was clearly in excess of 500 GW (Mulvaney, 2019); this made solar PVs the top source of new power in 2017 (Sawin et al., 2018). China has been one of the more aggressive countries in terms of installing solar power capacity, with about 154 GW of installed solar generating capacity as of June 2018 (Hill, 2018). China is not an isolated case, however – China had 33%, the USA had 13%, Japan had 12%, and India had 5% of the solar power capacity as of 2017 (Seetharaman, 2018). Globally, more solar PVs were installed in 2017 than all of the net additions of fossil fuels and nuclear power combined (Zervos, 2018).

A key factor in this growing capacity is cost. The price of solar-generated electricity continues to decrease in most locations in the world. In the USA, solar power is very competitive, and a significant fraction of new generating capacity was solar in 2016 and 2017 (30% in 2017; Perea et al, 2018). In fact, electric power generation from solar, wind, and natural gas were the lowest-cost technologies in 2018 (Rhodes et al., 2018; Lazard, 2018; Solar, 2018). Electricity-generation costs are in the range of $0.036 to $0.044/kWh in the USA for utility-size solar installations (Lazard, 2018). Residential costs are higher in most cases but, because of government incentives, there are many rooftop solar installations that are also economically advantageous. When homeowners put solar panels on their homes and have the benefit of free labor, for example, they often save on their electricity costs compared to the cost of purchasing electricity from the utility. These savings and costs associated with solar-generated electricity, naturally, will vary depending on global location, and will be best where there is excellent solar radiation, such as in Arizona, India, and Africa (Labordena, 2018).

One of the expectations is that the cost of solar-generated electricity will continue to decrease. The cost of solar panels in dollars per watt ($/W) has followed Swanson's Law, which predicts that the cost of solar power capacity in $/W will decrease by 20% for every doubling of the cumulative shipped volume of solar panels (Solar, 2018). In practical terms, this means that the prices of solar panels in $/W have been decreasing by about 10% per year (Solar, 2018). A lot of these cost decreases are because of improved efficiency, as well as decreased cost of production. Other cost decreases have been due to new technological developments, such as Passivated Emitter Rear Cell technology, which have improved conversion efficiency, with values as high as 23.6% being reported (Sawin et al., 2018). New materials also have the potential to lead to still lower costs. Weaver (2018) predicts that solar panel prices will decrease from $0.37/watt in 2017–2018 to $0.24/watt in 2022. That predicted decrease follows the 10% per year formula.

A final cost consideration is that, in a growing number of locations, rooftop solar installations also include battery storage. With battery storage, solar energy generated during the day can be stored until it is needed. Storage of energy is an important consideration because of the daily production cycle for solar-generated electricity. The full cost of electricity when it is supplied from batteries after storage should include the cost of storage. One method for estimating this full cost of electricity has been proposed (Lai and McCulloch, 2017). Naturally, one cost is purchasing the batteries, and also some electricity is lost while moving it into the batteries and bringing it out to be used – both of these factors should be included in the cost estimate. As noted in the previous chapter, though, the prices of batteries are dropping and this will alter the storage costs.

For some locations, there are less quantifiable costs and benefits to consider. In many countries, there are opportunities to use solar power generation to produce electricity where no grid exists. The outback in Australia has locations where solar-generated electric stations can be constructed and operated to serve one or more families in a small community. Africa has many rural locations where there is presently no electrical grid, and solar-generated electricity can be installed and used effectively there (Herscowitz, 2018; Labordena, 2018; Ochieng and Abiru, 2018). There is a need to bring electricity to many homes in Africa, and solar can do this economically and quickly. This idea applies even down to the level of solar lanterns, which can be put in the sun to charge batteries during the day and then used for lighting in the evening. Solar-generated electricity is often cost effective, even at small scale. Off-grid solar installations are already common in many remote areas of the world where an electrical grid is not available. Because transmission and distribution costs are about 40% of the cost of electricity (Solar, 2018), these costs are eliminated when electricity is generated on the roof of or next to a home.

Once the economics of solar power are favorable, where should the panels physically be placed? Solar panels can be installed on the rooftops of buildings, in parking lots, in deserts, over roads, and even to provide shade in parks. On the one hand, solar panels in locations such as parks and parking lots often can provide shade that is often beneficial. On the other hand, locations such as flat industrial roofs are practically invisible to a casual observer on the ground and therefore are unobtrusive and opaque changes. In California, there is a new requirement that all new homes must include solar power, starting in 2020 (Roselund and Weaver, 2018), and in some parts of Australia, more than 25% of homes have rooftop solar panels (Sawin et al., 2018). Where there are large numbers of solar panels, a significant fraction of the electricity in a region can be generated by solar. Furthermore, a person's perception that their neighbors are moving to solar power (e.g. from seeing the solar panels) can actually increase the uptake of solar power (Graziano and Gillingham, 2015).

The generation of solar power is not without some issues. Finding locations for large, utility-scale solar farms has sometimes led to controversies when there are associated environmental issues (Mulvaney, 2019). Of course, careful consideration of this and other issues may be needed for large projects. Also, solar panels can be manufactured from many different materials, and some of these materials entail significant environmental, health, and safety issues. This is an issue that should be addressed in the manufacturing process and in the end-of-life recycling of these materials. Mulvaney (2019) has some significant discussions about these issues in his book.

Finally, this section has noted several national and regional policy actions (e.g. in California) that have been helpful in accelerating shifts towards renewable energy, particularly solar power. The inverse is also true for certain other policies, such as tariffs. Presently, China is the country with the largest production of solar panels. Although China has many locations where solar panels have been installed, the country also sells solar panels to many countries around the world. One of the decisions that has been made by the United States is to place a tariff on Chinese solar panels in order to encourage local production. This is also an issue in India because of the benefits to the local economy of having domestic production of solar panels (Seetharaman, 2018): India's targets are to have 40% renewable energy by 2030 and 75% by 2050. In 2017, 40% of newly added electric generating capacity in India was solar (Lo, 2018).

5.3 Wind Energy

There has been great progress over the last 20 years in the generation of electricity from wind. The price of electricity from wind has been decreasing, and the size of new windmills has been increasing. The costs of wind power generated at sites on land depend on the wind resources at the site, but generation costs are now in the range of $0.029 to $0.063/kWh (Lazard, 2018; Sawin et al., 2018; Solar, 2018; Zervos, 2018). There are bid prices for energy close to $0.03/kWh in markets as diverse as Canada, India, Mexico, and Morocco (Zervos, 2018).

In 2017, 52 GW of new wind power-generating capacity was added, and global cumulative capacity is now at about 539 GW (Sawin et al.,

2018). Asia has been the largest regional market for wind power, with China being the largest market within that (about 188 GW at the end of 2017). The USA is second in terms of wind generating capacity, with 89 GW at the end of 2017. Globally, more than 5% of electric power generation is from wind. At least 13 countries in Europe, Central America, and South America met 10% or more of their electricity needs with wind power during 2017 (Zervos, 2018).

Compared to solar power, the productivity of wind power is more dependent on local conditions, but wind power is now the lowest-cost option for new power-generating capacity in a large number of markets (Zervos, 2018). For example, a University of Texas Energy Institute report shows that large parts of the Midwest United States have very suitable wind resources, and wind-generated electricity is the lowest-cost alternative (Rhodes et al., 2018). As the technology advances in wind-generated energy, the regions where it is feasible can be expected to expand, and as costs continue to decrease, wind power will become increasingly attractive. Thus, the technological progress in developing wind energy has significant value for society with respect to meeting the goals of reducing greenhouse gas emissions and improving air quality.

Approximately 96% of current wind power capacity is on land, but many countries, including the United Kingdom, Germany, China, and Belgium, have also developed offshore wind power (Sawin et al., 2018). The wind turbines are larger for offshore projects because they are not as constrained by the size limits that often exist for land sites. In 2017, the average wind turbine size in Europe was 5.9 MW for offshore installations; some are rated as high as 8 MW (Sawin et al., 2018) and several manufacturers have reported plans to introduce 10 MW and larger wind turbines (Zervos, 2018). The first offshore wind farm in the USA is off Block Island, Rhode Island, and the developer, Deepwater Wind, has announced that the Block Island farm is being purchased by Orsted, a Danish offshore wind company (ELP, 2018). The initial price of electricity on this wind farm was $0.244/kWh for this 30 MW project, which includes five wind turbines. While this price is relatively high, it is competitive with the past cost of electricity on Block Island, where fuel from petroleum sources has been used to generate electricity. In May 2018, Rhode Island announced that a 400 MW wind farm would be built by Deepwater Wind further offshore from the completed Block Island project (Kuffner, 2018).

The federal government has auctions to lease land for offshore wind sites off the coasts of Massachusetts, Rhode Island, California, and other states. There are plans for a large 800 MW project off the coast of Massachusetts. The estimated price for the generated electricity there is $0.065/kWh, which is competitive with other alternatives (Vyse, 2018). At the end of 2017, 17 countries had offshore wind generating electricity; the United Kingdom had the most offshore wind power capacity (at 6.8 GW). In 2017, the first commercial floating project was commissioned in Scotland (Zervos, 2018).

5.4 Renewable Energy Progress

In April 2019, the Federal Energy Regulatory Commission (FERC) reported that the electricity-generating capacity of renewables had exceeded that of coal for the first time (FERC, 2019). The figures were 21.56% for renewables and 21.55% for coal. The capacity for natural gas was 44.44% and for nuclear was 8.95%.

The electricity generated in April 2019 from renewables exceeded that from coal in the USA (Daley, 2019). In Texas, wind and solar generation was greater than electricity from coal for the first quarter of 2019 (Wamsted, 2019). One of the operational aspects is that electricity use is lower in April because heating and cooling use is less. Since there is no fuel cost for wind and solar, the reduction in operational output is from coal and natural gas plants (Wamsted, 2019).

For natural gas, the fraction of generating capacity (44.44%) is greater than the operational fraction (35%) because there are natural gas plants that are available to be used to provide peak power, and they are shut down for many days of the year. Since all of the electric power from wind and solar is used, the fraction of power from wind and solar is larger in the spring and fall. In summer, the natural gas plants are operated to meet peak power needs and coal plants are operated at a greater fraction of their capacity (Wamsted, 2019).

References

Daley (2019). For the first time, green power tops coal industry in energy production in April. Smithsonian; www.smithsonianmag.com/

ELP (2018, October 8). *Danish Offshore Wind Company Orsted Buys Deepwater Wind.* Electric Light and Power; www.elp.com/

FERC (2019). *Energy Infrastructure Update for April 2019.* Federal Energy Regulatory Commission; www.ferc.gov/

Graziano, M., and Gillingham, K. (2015). Spatial patterns of solar photovoltaic system adoption: The influence of neighbors and the built environment. *Journal of Economic Geography*, 15(4), 815–839; https://doi.org/10.1093/jeg/lbu036

Herscowitz, A. (2018, May 23). The unintended consequences of falling solar prices in Africa; https://medium.com/power-africa/

Hill, J.S. (2018, August 6). China installs 24.3 Gigawatts of solar in first half of 2018. Clean Technica; https://cleantechnica.com/

Jenniches, S. (2018). Assessing the regional economic impacts of renewable energy sources – A literature review. *Renewable and Sustainable Energy Reviews* 93: 35–51.

Kuffner, A. (2018, May 23). R.I. selects Deepwater Wind to build 400 MW offshore wind farm. *Providence Journal.*

Labordena, M. (2018, June 7). How sub-Saharan Africa can harness its big electricity opportunities. *The Conversation*; https://the conversation.com/

Lai, C.S. and McCulloch, M.D. (2017). Levelized cost of electricity for solar photovoltaic and electrical energy storage. *Applied Energy* 190: 191–203.

Lazard (2018, November 8). Lazard's levelized cost of energy and levelized cost of storage 2018. *Lazard*; www.lazard.com/

Lo, C. (2018, April 3). The road to 100 GW: Lighting up India with solar power. *Power Technology*; www.power-technology.com/

Mulvaney, D. (2019). *Solar Power*. University of California Press, Oakland, CA.

Ochieng, A. and Abiru, F. (2018, August 23). Solar startups are plugging Africa's energy gap. *Clean Technica*; https://cleantechnica.com/

Perea, A., Honeyman, C., Smith, C. et al. (2018). *U.S. Solar Market Insight: 2017 Year in Review*. Solar Energy Industries Association; www.greentechmedia.com/

Rhodes, J.D., King, C.W., Gulen, G. et al. (2018). *New U.S. Power Costs by County, with Environmental Externalities*. University of Texas at Austin; energy.utexas.edu/

Roselund, C. and Weaver, J. (2018, December 21). 2018 solar power year in review (Part 2). *PV Magazine*; https://pv-magazine-usa.com/

Sawin, J.L., Rutovitz, J., Sverrisson, F. et al. (2018). *Renewables 2018: Global Status Report*. Renewable Energy Policy Network; www.ren21.net/

Seetharaman, G. (2018, September 16). The road blocks in India's efforts to be world's greatest solar energy success story. *The Economic Times*.

Solar (2018). *Solar Electricity Costs*; http://solarcellcentral.com/

Vyse, G. (2018, November). *Offshore Wind Could be the Next Big Breakthrough in Renewable Energy*. Governing States and Localities; www.governing.com/

Wamsted (2019). IEEFC U.S.: April is shaping up to be momentous in transition from coal to renewables. Institute for Energy, Economics and Financial Analysis; ieefa.org/

Weaver, J. (2018, May 25). The path to $0.015/kWh solar power, and lower. *PV Magazine*; www.pv-magazine.com/

Zervos, A. (2018). *Renewables 2018: Global Status Report*. Renewable Energy Policy Network for the 21st Century; www.ren21.net/

6

Batteries

With contributions from Kate Preston

Abstract

Batteries are extremely important for electrical vehicles (EVs) and for the storage of electricity generated by wind and solar power that is to be delivered using the electrical grid. Because annual sales of batteries are projected to increase to more than $100 billion per year, there are many companies involved in research and development of new batteries. Most batteries for EVs today are lithium-ion batteries. With so many research groups, however, there is ongoing work regarding different materials that can be used for batteries, as well as different battery designs. The prices of batteries for EVs have decreased from more than $1000/kWh in 2010 to less than $200/kWh in 2018, and further decreases are expected. There are many challenges in developing batteries, including the effects of temperature, repeated charging and discharging, energy density, cost of materials, power flow rate, and more.

6.1 Introduction

As awareness of the effects of greenhouse gases increases, the push for a more sustainable way of life grows. Transportation is known to be one of the leading causes of climate change, with 23% of greenhouse gases coming from transportation-related emissions (Iclodean et al., 2017). To combat this increase in greenhouse gas emissions, the 'Paris Declaration on Electro-Mobility and Climate Change and Call to Action' was adopted in 2015, with the main goal being to reduce global warming (Iclodean, et. al, 2017). An important aspect of electrifying transportation is the batteries used in electric vehicles. Batteries can be beneficial not only to power the vehicle, but also to store energy from the grid in times when there may be surplus electricity generation. Research is currently being done to reduce the cost of batteries, extend battery life span, manage electric vehicle (EV) charging with the grid, and develop the most

TABLE 6.1

Average price of battery packs for electric vehicles, 2010–2018.

Year	Price ($/kWh)	Percent Change (%)
2010	1160	–
2011	899	22.5
2012	707	21.4
2013	650	8.1
2014	577	11.3
2015	373	35.4
2016	255	22.8
2017	214	25.7
2018	176	17.8

Source: Richter (2019).

environmentally friendly methods of battery production and disposal. The decline in battery prices is shown in Table 6.1 (Richter, 2019).

Research and development of batteries has been a very active area of investigation because of the significant value associated with batteries for EVs and storage of energy for the electrical grid. The annual sales of batteries have the potential to increase from $13 billion in 2017 to about $100 billion/year because of the global potential for EVs and energy storage in support of wind and solar energy generated for the grid. Recent data shows that the average cost for EV batteries was about $176/kWh in 2018 (McCrone, 2019). Current EVs have as much as 50–70 kWh of energy storage for cars that are to be used for longer-distance travel. Electric buses have about 300 kWh of battery storage.

There is the potential for prices to decrease to $50/kWh or less and for energy density to double or triple. Thus, by around 2050, there will likely be EVs with 100 kWh of battery storage and electric trucks and buses with 500 kWh of battery storage. At around that time there will also be more than 1 billion electric cars, more than 1 million electric buses, and more than 1 million large electric trucks in service. In addition, there may be of the order of 10 billion kWh of energy storage in support of the electrical grid, in addition to the EV batteries. There are challenges to address within all this progress, though. Along with the current issues and trends, this chapter reviews topics related to batteries, such as disposal of used batteries, the relationship between batteries and the smart grid, and some of the future possibilities in battery technology.

6.2 Design/Types

All batteries contain both a positive cathode and a negative anode. Currently, the leading two types of batteries are nickel metal hydride

(NiMH) and lithium-ion (Li-Ion). NiMH batteries use $Ni(OH)_2$ in the positive electrode and metal hydride in the negative electrode, while Li-Ion batteries use a metal oxide containing lithium in the positive electrode and graphite in the negative electrode (Young, et.al, 2013). The US Department of Energy (2017) describes NiMH batteries as having been successfully used because they are safe, are capable of handling rough conditions, have a substantial life span, and have acceptable specific energy and specific power (specific energy is how many kilowatt hours [kWh] can be stored per kilogram of mass, whereas specific power is the maximum kilowatts per kilogram a battery can deliver [Young et. al, 2013]). However, a few drawbacks of NiMH batteries are that they are expensive, generate heat, have high self-discharge, and lose hydrogen. Li-Ion batteries are now emerging as serious competitors to NiMH batteries due to their decreasing costs, increasing reliability, and good life span. One way to measure battery cost is the cost per kWh of energy. One goal is to lower the cost to $100/kWh to be more competitive with the purchase price of current combustion engine vehicles (Chediak, 2017). It is projected that this goal is attainable, likely with Li-Ion batteries, by 2020 (Chediak, 2017; McMahon, 2018; Morris, 2018).

Currently there is also significant interest in solid-state batteries because they have the potential to increase energy density. If battery energy density can be doubled, it will be very beneficial for EVs because the range of a vehicle could be increased to about 600 km (373 miles). Many investigators are working on solid-state batteries, with estimated global expenditures of more than $500 million/year in 2018 (Jaffe, 2019). There has been significant progress in developing solid-state batteries, and there are several different materials that are of interest (Zheng et al., 2018; Xin et al., 2017). By 2030, there may be commercial solid-state batteries in EVs that can be purchased (Gilboy, 2018).

There are several issues that are important in battery development, including energy density, durability, impact of temperature on battery service life, time to charge, operational features, and cost. Developing a commercial battery that meets all of the requirements for long-term use takes time, and important issues must be resolved successfully.

The design of batteries for grid energy storage can be different from those designed for EVs, particularly in terms of energy density. Energy density is more important for EVs, but systems with more volume or weight are unproblematically able to be used for grid storage. Research to improve the anode is being carried out by a number of research groups (Ball, 2019). There are a number of new battery designs that have promise; however, they need to be taken through the final stages of development and then produced commercially. The cost of production has been decreasing and the new battery needs to be competitive with the other alternatives at the time the new battery factory is ready for production. There is global competition that must be considered and trade agreements may be important. If prices of storage batteries can be reduced to less than $50/kWh, they will have greater value for grid energy storage applications, and sales volume will be significantly greater. Battery manufacturing capacity in 2018 was more than 100 GWh per year for EVs and electricity storage, with projections of about 500 GWh in 2022 (Ball, 2019).

Efforts for battery production are increasing. Tesla built a Gigafactory in Nevada with a goal to produce 35 GWh/year of lithium-ion cells in 2018 (Lambert, 2018a). China is advancing in the battery production market and now has over 140 EV battery manufacturers. By 2020, China's cell production global market share is expected to increase up to 70% (Perkowski, 2017). Japan and South Korea have major manufacturing plants (Lutsey et al., 2018). Many companies are working to make the battery making process as sustainable as possible. One major factor in this is recycling the materials, such as nickel, cobalt, and iron, from old batteries ('Battery Recycling', 2015). Volkswagen is planning to make a $1.1 billion investment for a new battery factory in Germany (Evarts, 2019a).

6.3 Battery Life

Battery life can be discussed in terms of two aspects: 1) the amount of time a battery holds a charge, and 2) the number of charge/discharge cycles a battery can endure. Both aspects of the battery's life can be affected by external factors such as temperature or rates of charge and discharge.

Temperature can affect battery life in multiple ways. Most obviously, in hot or cold weather, a portion of the battery's energy must be used to cool or heat the passenger cabin in an EV. Extreme temperatures decrease the range because the energy that would have been used to power the vehicle in moderate climates is now being used to control cabin temperatures, making it necessary for the vehicle to be charged after a shorter distance. High or low temperatures also affect the battery's performance more directly. In cold temperatures, batteries have a higher internal resistance and slower electrochemical processes, which cause the vehicle to accelerate more slowly (Pesaran et al., 2003). High temperatures can be harmful to batteries because the high temperatures cause more chemical activity, which speeds up battery degradation and self-discharge (Lindgren and Peter, 2016). Temperature can affect both the battery's ability to hold a charge and the number of cycles the battery can handle, especially at extreme highs and lows. It is also important to note that differences in individual cell temperatures can shorten the calendar life of a battery. According to Yang et al., the range of battery life is "between 5.2 years in Florida and 13.3 years in Alaska" (Yang et al., 2018). When cells are operating at different temperatures, the cells at higher temperatures tend to supply more power, which causes uneven degradation of the cells, which leads to imbalances in the pack and can cause warranty issues (Pesaran, et. al, 2003). To combat these issues, complete battery systems include a battery thermal management system (BTMS). Generic BTMSs are comprised of cooling, heating, and insulation elements that are adjusted to regulate each specific battery. They are used not only to enhance the life span of the battery, but also to keep the battery operating safely (Khan et. al, 2017). A simple way in which EV owners can mitigate these temperature imbalance issues is to park an EV in a location that helps to keep the battery cooler. For example, a garage or other sheltered spot will help to prevent heating from the ambient temperature on hot days. A more advanced option is to work towards the further building of solar-powered charging stations (SPCSs), which help keep the battery temperature down on a hot day by providing shade (i.e. the solar panels) while also charging the battery.

The charging of a battery impacts its life span. An electrolyte between the two electrodes conducts current, and conduction occurs when ions of the electrodes are exchanged through the electrolyte. The exchange of the ions occurs at the solid electrolyte interface (SEI) and forms an SEI layer. Material builds up at this layer, creating internal resistance until the resistance is so strong that ions can no longer be transferred, and the battery has reached the end of its calendar life (Arcus, 2016). At high temperatures and at high voltage, this process happens faster. Charging can occur at three levels: Level 1, level 2, and level 3, or DC fast charging. As the level increases, so does the voltage, and the amount of time it takes to charge the vehicle decreases. The faster the charge, the faster the charging will degrade the battery; therefore, level 1 charging is healthiest for the battery and level 3 is the least healthy. A test done on four BEVs, where two were charged using DC fast charging and the other two were charged using level 2, showed a small but noticeable difference, with greater battery capacity lost through DC charging (Shirk and Wishart, 2015).

Batteries are equipped with a battery management system (BMS) that ensures the battery is running safely and helps to protect it. Each charge and discharge has an impact on battery life span. When a battery cell is fully charged, it is at 100% state of charge (SoC). When a battery is fully discharged, it is said to be at 100% depth of discharge (DoD) (Arcus, 2016). Consistently taking a battery to 100% SoC or DoD is not good for the health of the battery, so a BMS prevents the battery from reaching those points. The BMS also "ensures that the energy of the battery is optimized to power the product" and "that the risk of damaging the battery is minimal" (Hu, 2012).

Lastly, it should be remembered that maintenance costs on EVs are lower because the battery and motor have fewer "wear-and-tear" parts, compared to a conventional, internal combustion engine. Electric vehicles lack many of the periodic maintenance items associated with conventional engines, such as spark plugs, valves, belts, hoses, and catalytic converters (Gorzelany, 2019). In fact, there are entire systems – the carburetor/fuel injection system, the timing system, the transmission, and the exhaust system – that are basically no longer relevant or part of an EV.

6.4 Charging

As mentioned earlier, EV charging can be done at three main levels. Level 1 charging requires a 120-volt outlet, which is typical for a household outlet and can add between two and five miles of range per hour of charging. Level 2 charging requires a 240-volt outlet and provides 10–25 miles of range per hour. Level 3 DC fast charging provides an 80% charge in around 30 minutes (Schaal, 2018). For a Honda Clarity Electric, for instance, getting a full charge using level 1 charging takes 19 hours, level 2 charging takes 3.5 hours, and an 80% charge can be achieved in 30 minutes when using DC fast charging ('Clarity FAQs', 2017). It is important to note, however, that different EVs have different power "acceptance rates" and charging stations have varying power delivery ratings. If an EV has a higher acceptance rate than the charging station's delivery rating, then the time it takes to charge will be limited by the charging station. If a charging station is rated for higher power delivery than an EV's acceptance rate, the amount of time needed for charging will be determined by the EV (Guinn, 2018).

Many new developments in technology aimed at making EV charging more convenient for drivers are arising. For example, Electrify America launched a mobile app that allows users to locate a charger, pay for the charge, and track the charging session as the EV is charging (Electrify America LLC, 2019). Another app called "ampUp – reserved EV charging" allows users to reserve charging stations in advance, which can be very beneficial for long-distance trips where users need to know when and where they can recharge along the way (ampUp). ChargeHub has a similar app that even allows users to rate their charging experience and leave comments for other users to see (ChargeHub, 2019).

There are several secondary issues related to EV battery charging, in addition to the basic issue of plugging in and taking in electricity. One of the secondary issues was mentioned earlier: EVs typically have a thermal management system, and it is important to note that this system uses energy. If the thermal management system has to cool the battery the entire time that the EV is parked in the hot sun, some of the EV's charge will be drained (Connor, 2011). SPCSs combat this issue by providing shade for the EV so it can sit in a cooler environment during the day and use solar energy for charging. For example, Envision makes SPCSs for parking lots that include solar panels with sun-tracking technology, which is said to increase electricity production 18–25% (Motavalli, 2010).

Another secondary issue that can influence charging is the fact that most EVs get some charge while driving, through the process of regenerative braking. For regenerative braking, instead of acting as an electric motor, the motor works as a generator and converts the kinetic energy of the vehicle's motion into electrical energy that can charge the battery (Young, et. al., 2013). The amount of energy recovered is dependent upon the driving conditions. Outside of the car, the aerodynamic loss, friction of the tires, and high speeds help to slow the car but do not create recoverable energy (Solberg, 2007).

Yet another secondary issue is the development of wireless charging, which is already on the market for electric buses. Widespread wireless charging is still in development, but Momentum Dynamics has a "200-kW wireless charging system" that they claim is more energy efficient than current plug-in chargers (Lambert, 2018b).

6.5 Battery Disposal

After about seven to ten years, EV batteries can no longer effectively power the vehicle and must be replaced (Gibson, 2018), leaving the question of what to do with the old batteries. Today's batteries contain toxic chemicals that would be environmentally hazardous if they were to end up in landfills. On the other hand, batteries contain valuable materials that could be recovered and reused, minimizing the need to obtain new resources (Engel, 2016). For these reasons, China and the European Union require EV owners to recycle their batteries, and it is predicted that similar laws will be implemented soon in the USA (Gibson, 2018). Fortunately, EV drivers today have the option to reuse or recycle their used batteries.

There are many reasons why users should recycle their batteries. For one, the sheer size and weight of an EV battery makes recycling an easier and more practical option than throwing them away (Field, 2018). Because a continuous goal for EV batteries is to make them cheaper and lighter, they will likely decrease in size, use less material, be less expensive, and be lighter when the time comes to replace them. If an owner chooses to do so, they will almost certainly be able to upgrade the properties of their car when getting a new battery.

It is also important to continue recycling because the recycling of lithium-ion batteries can have a number of social, economic, and environmental benefits (Nikolewski, 2018). To take one aspect, the recycling of lithium-ion batteries could significantly decrease the need for new cobalt and simultaneously reduce the cost of batteries (Nikolewski, 2018). More than half the world's cobalt supply is contained within the Democratic Republic of the Congo. This country has endured two civil wars and has also faced assertions of using child labor to extract cobalt (Nikolewski, 2018).

Another important option is the reusing or repurposing of EV batteries. Many manufacturers are already exploring reuse options. Nissan has powered streetlights using old batteries in Japan, Renault is using batteries to back up elevators in Paris, and GM is using old Chevy Volt batteries to back up its data center in Michigan (Gibson, 2018). Electric and hybrid batteries can also be useful in a household when paired with solar panels to store excess electricity produced during the day ('Battery Recycling', 2015). Reused batteries can also be useful as electric bike batteries, which is a technology that Powervault and Aceleron have created (Gibson, 2018). Nissan has also created the Nissan x OPUS – an inflatable camping trailer that runs using old Nissan Leaf batteries (Evarts, 2019b).

6.6 Batteries and the Power Grid

As the number of EVs on the road increases, it also means an increase in total demand for electricity. An important part of the EV infrastructure is making sure the grid can support the charging of EVs in use. Many different methods are being used and tested to make EV integration with the grid go as smoothly as possible. One of these methods is known as "time-of-use rates" or "time-of-use prices". The idea is to incentivize drivers to charge their vehicles at times when the least stress is put on the grid. For instance, Portland General Electric implemented a time-of-use rate plan in its territory in Oregon: On-peak charging costs 19.9 cents/kWh, midpeak costs 14.5 cents/kWh and off-peak is 4.2 cents/kWh (Cohn, 2018). Residential users of Pacific Gas & Electric could save up to $70 annually by running their dishwasher at 9pm instead of 6pm (assuming they run a 1.5-kW dishwasher 20 times a month on PG&E's E-6 TOU rate) (Fields, 2018).

EVs can also be used as "mobile storage". According to Gretchen Bakke, using EVs as storage would allow the energy demand to move around as people do (Bakke, 2017; Davar, 2017). This may be particularly useful in off-grid environments. If an EV owner's workplace offers charging during the day and the EV contains a significant battery, the vehicle could be charged while at work and then later hooked up to the off-grid home. This would then serve as a power supply for cooking, appliances, or to do laundry (Erickson, 2019). This could also be useful for helping energy supply to be more in line with demand. Davar summarized this situation by noting that "the sun clocks out for the day when everyone gets home and turns on all sorts of energy-consuming devices", which is why using stored energy in an EV battery would be valuable (Davar, 2017).

6.7 The Future of EV Batteries

The future of EV batteries looks increasingly promising as sales continue to rise dramatically and prices continue to drop. One Chinese company already claims that in 2020 it will be able to make battery packs for under $100/kWh, which is a typical calculated price for EVs to be cheaper than combustion engine vehicles (Evarts, 2018). Bloomberg New Energy Finance projects that lithium battery packs will cost $70 kWh in 2030 (Morsy, 2018), and Lei Zheng, Envision CEO, has predicted $50/kWh battery prices by 2025 (Rufiange, 2018). Table 6.1 shows that the average decline in battery prices has historically been about 20% each year. If this continues, battery prices will be less than $50/kWh in 2025. This is not taking into consideration the possibility of "transformational" developments, such as new battery types emerging. For example, research is currently being done to develop zinc-air and other metal-air batteries. Zinc-air batteries have captured the focus of research due to their potential for high energy densities and low production costs (Li and Dai, 2014). Electrify America announced, along with the launch of their app, that the company now offers two membership plans for EV drivers and a pricing structure that lowers the introductory starting cost by about 20% (Electrify America LLC, 2019).

Lastly, other factors pushing the growth of EV batteries continue to emerge and develop. The California Legislature is pushing for carbon-free power by 2045 (Szymkowski, 2018). Many countries have publicly released their goals to decrease or eliminate internal combustion engine vehicle sales by certain dates (Burch and Gilchrist, 2018). The significant increase in battery energy density has already allowed for ranges of above 200 miles on a full charge for many current EVs (Gorzelany, 2019).

References

ampUp. (n.d.); http://ampup.io/

Arcus, C. (2016, May 31). Battery lifetime: How long can electric vehicle batteries last? *CleanTechnica*; cleantechnica.com/2016/05/31/battery-lifetime-long-can-electric-vehicle-batteries-last/

Bakke, G.A. (2017). *The Grid: The Fraying Wires Between Americans and Our Energy Future*. New York, NY, Bloomsbury.

Ball, J. (2019, June). The race to build a better battery. *Fortune* 202–215; fortune.com/

Insteading (2015). Battery recycling: How we'll keep electric vehicle batteries out of landfills; https://insteading.com/blog/battery-recycling-how-well-keep-electric-vehicle-batteries-out-of-landfills/

Burch, I. and Gilchrist, J. (2018). Survey of global activity to phase out internal combustion engine vehicles [revision]. Center for Climate Change Protection. Retrieved May 31, 2019; https://climateprotection.org/wp-content/uploads/2018/10/Survey-on-Global-Activities-to-Phase-Out-ICE-Vehicles-FINAL-Oct-3-2018.pdf

ChargeHub (2019). Definitive guide on how to charge an electric car; https://chargehub.com/en/electric-car-charging-guide.html#publiccharging

Chediak, M. (2017, December 5). The latest bull case for electric cars: The cheapest batteries ever. *Bloomberg.com*; www.bloomberg.com/news/articles/2017-12-05/latest-bull-case-for-electric-cars-the-cheapest-batteries-ever

Clarity FAQs (2017). Retrieved April 16, 2018; https://automobiles.honda.com/clarity-electric/landing-pages/clarity-faq

Cohn, L. (2018). Postcard from the future: Latest on how states are preparing the grid for electric vehicles. *Microgrid Knowledge newsletter*; https://microgridknowledge.com/ev-infrastructure-states/

Connor, P. (2011). Eight tips to extend battery life of your electric car; www.plugincars.com/eight-tips-extend-battery-life-your-electric-car-107938.html

Davar, Z. (2017). How electric vehicles can drive renewable energy forward. Medium. Retrieved April 04, 2019; https://medium.com/cleantech-rising/how-electric-vehicles-can-drive-renewable-energy-forward-485b8e792db6

Electrify America LLC (2019). Electrify America launches mobile app to enhance electric vehicle charging experience [press release pdf]. Retrieved May 16, 2019; www.electrifyamerica.com/news-updates

Engel, J. (2016). Development perspectives of lithium-ion recycling processes for electric vehicle batteries. Open Access Master's Theses. Paper 905. http://digitalcommons.uri.edu/theses/905

Erickson, L.E. (2019, April 11). Personal communication.

Evarts, E.C. (2018). Electric cars cheaper than gas ones by 2020, says Chinese battery supplier;www.greencarreports.com/news/1120318_electric-cars-cheaper-than-gas-ones-by-2020-says-chinese-battery-supplier

Evarts, E.C. (2019a). Mercedes-Benz EQC, VW's battery factory, Hyundai's electric sports car: Today's Car News; www2.greencarreports.com/news/1123106_mercedes-benz-eqc-vws-battery-factory-hyundais-electric-sports-car-todays-car-news

Evarts, E.C. (2019b). Reused Nissan Leaf batteries make happy campers with powered pop-up trailer; www.greencarreports.com/news/1121660_reused-nissan-leaf-batteries-make-happy-campers-with-powered-pop-up-trailer

Field, K. (2018). Yes, Tesla recycles all of its spent batteries & wants to do more in the future [web log comment]; https://cleantechnica.com/2018/06/07/yes-tesla-recycles-all-of-its-spent-batteries-wants-to-do-more-in-the-future/

Fields, S. (2018). What are time of use rates? How do they work? EnergySage. Retrieved June 1, 2019; https://news.energysage.com/understanding-time-of-use-rates/

Gibson, R. (2018). How electric vehicle batteries are reused or recycled [web log post]; www.fleetcarma.com/electric-vehicle-batteries-reused-recycled/

Gilboy, J. (2018, November 14). *Panasonic North American CEO: Solid-state EV batteries at least a decade away*. The Drive; www.thedrive.com/news/

Gorzelany, J. (2019). 10 reasons to buy an electric vehicle in 2019. Retrieved May 31, 2019; www.myev.com/research/why-buy/10-reasons-to-buy-an-electric-vehicle-in-2019

Guinn, S. (2018). Level 1 vs Level 2 electric vehicle charging stations. Retrieved April 16, 2018; www.clippercreek.com/level-1-level-2-charging-stations/

Hu, R. (2012). *Battery Management System For Electric Vehicle Applications* [unpublished doctoral dissertation]. University of Windsor.

Iclodean, C., Varga, B., Burnete, N. et. al. (2017). Comparison of different battery types for electric vehicles. *IOP Conf. Ser.: Mater. Sci. Eng.* 252: 012058.

Jaffe, M. (2019, January 2). How a Louisville company aims to make electric cars cost less and drive farther using technology from CU. *The Colorado Sun*; coloradosun.com/

Khan, M.R., Swierczynski, M.J., and Kær, S.K. (2017). *Towards an Ultimate Battery Thermal Management System*. Multidisciplinary Digital Publishing Institute (3), 9th ser., 1–18. Retrieved March 15, 2018.

Lambert, F. (2018a). Tesla increases hiring effort at Gigafactory 1 to reach goal of 35 GWh of battery production. Retrieved March 14, 2018; https://electrek.co/2018/01/03/tesla-gigafactory-hiring-effort-battery-production/

Lambert, F. (2018b). The first 200-kW wireless charging system for electric buses is deployed; https://electrek.co/2018/04/19/200-kw-wireless-charging-system-electric-buses/

Lindgren, J. and Peter, L. D. (2016). Effect of extreme temperatures on battery charging and performance of electric vehicles. Retrieved February 28, 2018; www.sciencedirect.com/science/article/pii/S0378775316308941

Li, Y. and Dai, H. (2014). Recent advances in zinc-air batteries. *Chemical Society Reviews* 43: 5257–5275.

Lutsey, N., Grant, M., Wappelhorst, S., and Zhou, H. (2018, May). Power play: How governments are spurring the electric vehicle industry. International Council on Clean Transportation White Paper; www.the icct.org/

McCrone, A. (2019). Transition in Energy, Transport – 10 Predictions for 2019. *BloombergNEF*; about.bnef.com/

McMahon, J. (2018, December 4). Chinese company says it will soon cross $100 battery threshold, slaying the gasoline car. *Forbes*; www.forbes.com/

Morris, C. (2018, June 28). Tesla is approaching the magic battery cost number. *Inside EVs*; insideevs.com/

Morsy, S. (2018). Electric vehicles; https://bnef.turtl.co/story/evo2018?teaser=true

Motavalli, J. (2010, October 18). A perfect marriage: Electric car charging and solar power; www.mnn.com/green-tech/transportation/blogs/a-perfect-marriage-electric-car-charging-and-solar-power

Nikolewski, R. (2018, March 16). New way to recycle lithium-ion batteries could be a lifeline for electric cars and the environment. *Los Angeles Times*; www.latimes.com/business/technology/la-fi-lithium-ion-battery-recycling-20180316-story.html

Perkowski, J. (2017, August 3). EV batteries: A $240 billion industry in the making that China wants to take charge of. Retrieved April 11, 2018; www.forbes.com/sites/jackperkowski/2017/08/03/ev-batteries-a-240-billion-industry-in-the-making/#743918173f08

Pesaran, A., Vlahinos, A., and Stuart, T. (2003). Cooling and preheating of batteries in hybrid electric vehicles. *Japan Society Mechanical Engineers*, 1–7. Retrieved March 15, 2018; www.nrel.gov/transportation/assets/pdfs/jte_2003-633_sw_ap.pdf

Richter, F. (2019, March 12). Can falling battery prices push electric vehicles? *The Statistics Portal*; www.statista.com

Rufiange, D. (2018, December 20). Electric cars cheaper than gas-fed vehicles by 2025. *Auto123.com*; auto123.com/

Schaal, E. (2018, March 2). A simple guide to electric vehicle charging. Retrieved March 29, 2018; www.fleetcarma.com/electric-vehicle-charging-guide/

Shirk, M. and Wishart, J. (2015, March). Effects of electric vehicle fast charging on battery life and vehicle performance. SAE 2015 World Congress, Detroit, MI; www.osti.gov/servlets/purl/1236824.

Solberg, G. (2007, June 29). The magic of Tesla Roadster regenerative braking. Retrieved May 3, 2018; www.tesla.com/blog/magic-tesla-roadster-regenerative-braking

Szymkowski, S. (2018, September 10). Leaf/Outlander PHEV sales race, California goes zero-carbon by 2045, and electric car poll: Today's Car News; www2.greencarreports.com/news/1118688_leaf-outlander-phev-sales-race-california-goes-zero-carbon-by-2045-and-electric-car-poll-todays-car-news

US Department of Energy (2017). *Batteries for Hybrid and Plug-in Vehicles*. Alternate Fuels Data Center; www.afdc.energy.gov/

Xin, S., You, Y., Wang, S. et al. (2017). Solid-state lithium metal batteries promoted by nanotechnology: Progress and prospects. *ACS Energy Letters*, doi:10.1021/acsenergylett.7b00175.

Yang, F., Xie, Y., Deng, Y., and Yuan, C. (2018). Predictive modeling of battery degradation and greenhouse gas emissions from U.S. state-level electric vehicle operation. *Nature Communications*, 9(1). doi:10.1038/s41467-018-04826-0.

Young, K., Wang, C., Wang, L.Y., and Strunz, K. (2013). Chapter 2: Electric Vehicle Battery Technologies. Rodrigo Garcia-Valle and João A. Peças Lopes (eds.), *Electric Vehicle Integration into Modern Power Networks*, © Springer Science + Business Media, New York.

Zheng, F., Kotobuki, M., and Song, S. et al. (2018). Review on solid electrolytes for all-solid lithium-ion batteries. *Journal of Power Sources* 389: 198–213.

7

Smart Grid

Abstract

The smart grid is a recent development that evolved as the marriage of our electronic communication abilities and the need for better grid management (balancing supply and demand for electricity across increasingly diverse sources and consumers). Managing the electrical grid now includes distributing electricity generated by wind and solar that flows into the grid as it is generated by the utility and others. Energy storage in batteries is also increasing, and this becomes an important operational option for the grid manager. Automation may be used to assist smart grid operators and customers with grid management. Adjustable (dynamic) prices can be used in a smart grid to provide incentives that reduce use at peak demand times and increase use at times of low demand but high supply. With the smart grid and dynamic prices, customers have the opportunity to reduce the cost of their electricity by making decisions that help balance supply and demand.

7.1 Introduction

One of the greatest logistical needs related to reducing greenhouse gas emissions through the use of wind and solar energy is the development of a smart grid. To understand why this matters so much, it is useful to first take stock of how our electrical grid is set up now. Historically, the electrical grid manager was the energy producer, and they were concerned with generating an appropriate amount of electricity, at the necessary times, to supply electricity sufficient for the demand. It is a fairly simple arrangement: Demand information flows one way, and energy supply flows back. One analogy would be a parent trying to care for the needs of a small child – the child signals what it needs and the parent provides it.

Because the wind blows when it happens to blow and the sun shines when it shines, electricity generation from wind and solar energy fluctuates over the course of hours in a day and across days. These fluctuations, of course, do not always match up with consumer demands for electrical power, so wind and solar energy complicates the role of the grid. Wind and solar energy will often need to be stored to manage supply and demand for electricity generation and distribution. On top of that, renewable energy generation – in particular solar energy – is increasingly also being generated by individual consumers and businesses. When these small-scale energy sources produce electricity in excess of their own needs, that electricity should ideally be put into the grid for others to use. The possible analogy is now adults interacting, in different amounts at different times, some with more resources (e.g. utility companies), but all to some extent being able to both supply and demand resources. All the while, this grid needs to manage overall supply and demand of electricity economically, with a greater focus on managing consistency relative to demand compared to the past. This needs to be a "smart" grid.

Correctly implemented, a smart grid is beneficial in the management of systems that have significant portions of their electric power generated by wind and solar energy (Hossain et al., 2016). The smart grid enables both grid operators and customers to reduce costs and improve efficiency by balancing supply and demand through better communication and pricing of electricity (Astarloa et al., 2017). Speer et al. (2015) adapted the following definition of a smart grid from the European Technology Platform Smart Grid (ETPSG):

> [The] smart grid is a concept and vision that captures a range of advanced information, sensing, communications, control, and energy technologies. Taken together, these result in an electric power system that can intelligently integrate the actions of all connected users – from power generators to electric consumers to those that both produce and consume electricity ('prosumers') – to efficiently deliver sustainable, economic, and secure electricity supplies.

7.2 Features of a Smart Grid

The smart grid is also a complicated grid; if it were simple, there would be no need for it to be smart. The smart grid includes the electric system that joins production sources, transmission lines, distribution systems, storage systems, and customers with smart meters. There is a communication system that includes computers, meters, information displays, and an Internet with current information and resources that may be needed. Computer programs with automated features, cybersecurity protection, and control systems are present and there may be an active human operator of the smart grid. The system is connected to a larger grid where electricity may be purchased or sold. The system allows for customer engagement and decision making to reduce costs by purchasing electricity at times when prices are low and avoiding some uses at peak power times.

People may be needed as part of the smart grid management, such as a smart grid operator and a distribution system operator, who has responsibility for the portion of the smart grid where power flows to meet electricity demands. Of course, there may also be consumer/producers with solar panels that generate excess electricity as part of the distribution system, and power may flow into the grid from multiple locations because of this. Additionally, there are already transmission system operators who are concerned with power flow from one part of the country to another. The smart grid for a community is similarly connected to a larger grid that has someone who manages it (Mwasilu et al., 2014).

In order to balance supply and demand, a smart grid should have the ability to use prices as incentives to alter demand. If the utility has excess supply and wishes to increase demand, they can lower the price and encourage, for example, electric vehicle (EV) owners to charge their cars by informing them of the lower price through communication on the smart grid information system. The utility may even have arrangements with EV owners who have parked in a parking lot with a solar-powered charging infrastructure that includes electric vehicle supply equipment (EVSE) to charge EVs when the price is sufficiently low. Excess supply may also be sold into the larger grid. Since the smart grid likely includes storage batteries, the utility can alternatively choose to store electricity in its storage batteries. The smart grid includes software and computer programs that help guide this and other decision making processes.

The smart grid must be able to distribute all the electric power that is generated and supply the power that is needed at all times. At peak power demand times, prices may be higher to discourage power use for charging EVs, drying clothes, and other uses that can be scheduled at alternative times when prices are lower. There are some places where utilities have optional contracts with a subset of users that allow power delivery to be briefly interrupted when demand is high. Stored energy in batteries can be used to supply power to the grid at peak power demand times, maximizing the value of that resource. This later option may include power flow from EV batteries into the grid, as explained later on. Many utilities have generating equipment that is powered by natural gas that can be started and used when even more power is needed. Finally, electric power can be purchased from the larger grid if it is available.

The smart grid, properly implemented, can improve safety and reliability. Through improved communication and incentives, it can achieve better efficiency, flexibility, and compatibility. Integration of the various parts of the grid is enhanced. The smart grid becomes increasingly important as the portion of energy from renewable sources increases and grid management becomes more complex (Munje, 2017). With the smart grid, consumers have more choices, and greater efficiency is possible because of prosumers and consumers working with others to modify supply and demand to have improved outcomes.

With smart meters and smart grids, it is possible to make use of computers, machine learning, and the Internet of Things to gather and process information that gives us a better understanding of the needs and habits of customers, so that better services can be provided (Reka and Dragicevic, 2018; SAS, 2019). Artificial intelligence technology can be used effectively with advanced network technology to make the smart grid efficient in processing data. The Internet of Things links objects to the Internet to create a global infrastructure for the information society, enabling communication that allows intelligent devices to work cooperatively for the benefit of society. For example, an intelligent system of the customer and the smart grid can decide when to turn on the electricity to charge the battery on an EV. Smart meters are able to provide data that can be used with machine learning to make better decisions that improve efficiency and security (SAS, 2016).

The two-way communication in smart grids allows for a significant increase in collected data and much greater consumer engagement and participation. Consumers can adjust their consumption of electricity and modify their behavior when they perform some tasks. Solar panels for electricity generation and storage behind the meter are becoming more common, and the smart grid and smart meters make the grid more robust and secure. This does require that all involved follow established communication standards and electrical system standards to provide an effective service (Bikmetov et al., 2017). Several different computing platforms and various web services are available for energy management and control of the smart grid. Internet of Things technologies may be used for automated management of smart grid decision making to achieve effective energy use. The benefits of the Internet of Things for the smart grid include better customer service, greater energy efficiency, improved data-driven decision making, increased integration of distributed energy resources, and better customer engagement (SAS, 2016). Machine learning, using data from smart meters and other information, may also lead to increased cybersecurity (e.g. by recognizing anomalies more quickly) and reductions in restoration time when problems are encountered.

The digital revolution associated with the Internet of Things and the process of developing automated decision systems that make use of machine learning have been described as the fourth industrial revolution (Marr, 2018; Reka and Dragicevic, 2018). The smart grid is one application of this new approach to improving communication and decision making. It is now possible to automate some smart grid applications so that higher-demand activities, such as EV charging, can be turned on when demand for electricity needs to be increased and there is a price incentive to do so.

7.3 Energy Storage

The smart grid, like any electrical grid, is fundamentally about balancing, at all times, the electricity that is being generated (coming in) and the electricity that is being demanded by customers (flowing out). There are more sources of generation and more directions of flow, but perhaps one of the qualitatively distinct aspects of the smart grid is the addition of storage and the management of the electrical flows to and from that storage.

With the flexibility of the smart grid, generation and storage of energy could be accomplished in many different ways, and coordinated so that the benefits and weaknesses of each are counterbalanced to some extent. The generation of renewable electricity may be via wind, solar, hydro, or even methane gas from anaerobic digestion. Non-renewable electricity may also still be used within the system. Similarly, the storage of energy can be accomplished using a variety of methods. Large utilities, with large power-generating capabilities, may have battery storage as part of their infrastructure, because storage is beneficial where a significant portion of the supply is from wind and solar energy sources (Munje, 2017). Although batteries are the most often considered option for large-scale storage, other options are also possible. For instance, potential energy can be stored in water reservoirs and used to generate electricity when needed. Water can be pumped back into a reservoir when excess electricity needs to be used, then released to produce hydroelectricity.

There is a lot of potential for the smart grid to use another method of electricity storage, as yet not utilized: EV batteries. EV batteries can be used for smart grid management of supply and demand (Tan et al., 2016). The potential energy storage capabilities of the millions of EVs already in use is immense. Many battery electric vehicles (BEVs) have batteries that can store more than 40 kWh. Furthermore, when a BEV is being used to travel locally, there is excess battery capacity which can be used to make the smart grid more efficient. These EVs, parked and plugged into the grid already, can be beneficial to the management of supply and demand of electricity if they are integrated into the smart grid system.

It would therefore be beneficial for smart grid management to have parking lots full of EVs that can be charged when the system operator needs to increase demand for renewable power that is being generated. This strategy – delivering power to charge EVs when they are coincidingly being used to help balance supply and demand – may be offered at a lower price to encourage EV owners to help the smart grid in this way. The smart grid management technology exists to have EV owners arrive at a parking lot, plug in their car to a charging station, and communicate their wishes with respect to battery charging. There may be automated systems that help to manage parking lots full of EVs based on the state of supply and demand of the smart grid and the desires of each EV owner. EV owners with large battery capacity may be able to charge their EV at times when the price is low and it is beneficial for the smart grid and/or utility. The reduced price can be offered because the grid efficiency is improved by charging EVs when it helps to balance supply and demand.

Managing the variability associated with wind and solar generation and the high power flows of BEV fast charging is easier with a smart grid that has good communication features and battery storage. The communication features allow customers to have a better understanding of how they can be beneficial to the balancing of supply and demand. Intelligent customer decision making is very beneficial in smart grid management. It may also involve an EV aggregator (either a person or a computer program) that manages a parking lot with solar panels, charging stations, and a microgrid, and that works to carry out the wishes of the EV owners who have parked and connected their EVs to charging stations.

7.4 Regulatory Topics

Electric utilities in many places may be a regulated monopoly that must work with a commission on rates and rate structure. Because of smart grid developments, renewable energy, and the rising prevalence of EVs, there is a need to transition to dynamic rate structures that have higher electricity prices at peak power times when demand is large. Some of this is not actually novel; parts of California developed and implemented time-of-use pricing over three decades ago. In other areas of life, price is naturally a reflection of supply and demand. For example, fruits and vegetables cost different amounts depending on the season and time of year; electricity can be the same (and, in both cases, you may be able to grow your own!). There is a need for those serving on regulatory commissions to find good ways to manage prices in ways that are equitable and that encourage efficient use of smart grids to balance supply and demand. The power generated by solar panels on customer homes and in parking lots must be appropriately valued so all of the electricity that is generated can be delivered to locations where it is needed or can be stored. Storage can include behind-the-meter storage by prosumers who both generate and use electricity.

7.5 Seasons and Demand

One of the smart grid issues is how to balance supply and demand across seasons. If renewable energy is used to generate most of the electricity, there may be challenges of balancing supply and demand in spring and fall, when demand for electricity is often lower because of reduced needs for heating or air conditioning. Some of this fluctuation may be offset by decreasing output from certain types of electricity generation, such as non-renewable energy power plants (much like how the current grid operates). It is also possible that there are industrial uses for this electricity that are less time sensitive and would benefit from lower seasonal prices. For example, low-cost electricity could be used for the production of hydrogen from water by electrolysis. Hydrogen is used as a raw material in the production of ammonia fertilizer, and spring and fall are times of high demand for fertilizer to be applied to crops.

7.6 Interface Issues

Lastly, there are several technological interface issues which need to be fully realized and correctly accommodated within the modern smart grid. The smart grid includes distributed generation and interfaces that must be managed to maintain the desired voltage and harmonics. Solar distributed generation may cause the voltage to rise when there is a reverse flow of electricity from the microgrid, with solar panels feeding electricity into the smart grid (Shaukat et al., 2018). Smart grid management of distribution systems with many distributed generators requires attention to be paid to voltage regulation and the balance of supply and demand. There also are communication interface issues that may reduce the quality of the communication. Security concerns must be managed to avoid problems such as corruption of the smart grid or improper entry into the grid. And, of course, reliability issues must be continuously managed to deliver a quality product at all times (Shaukat et al., 2018).

References

Astarloa, B., Kaakeh, A., Lombardi, M. et al. (2017). *The Future of Electricity: New Technologies Transforming the Grid Edge*. World Economic Forum, Geneva Switzerland.

Bikmetov, R., Raja, M.Y.A., Sane, T.U. (2017). *Infrastructure and Applications of Internet of Things in Smart Grids: A Survey*. North America Power Symposium, IEEE; https://ieeexplore.ieee.org/

Hossain, M.S., Madlool, N.A., Rahim, N.A. et al. (2016). Role of smart grid in renewable energy: An overview. *Renewable and Sustainable Energy Reviews* 60: 1168–1184.

Marr, B. (2018, August 13). The 4th industrial revolution is here – Are you ready? *Forbes*.

Munje, S. (2017). Renewable energy integration into smart grid-energy storage technologies and challenges. *International Research Journal of Engineering and technology* 4: 6, e-ISSN: 2395-0056.

Mwasilu, F., Justo, J.J., Kim, E.K. et al. (2014). Electric vehicle and smart grid interaction: A review on vehicle to grid and renewable energy sources interaction. *Renewable and Sustainable Energy Reviews* 34: 501–516.

Reka, S.S. and Dragicevic, T. (2018). Future effectual role of energy delivery: A comprehensive review of Internet of Things and smart grid. *Renewable and Sustainable Energy Reviews* 91: 90–108.

SAS (2019). *A Non-Geek's A to Z Guide to the Internet of Things*. SAS Institute, Cary, NC.

SAS (2016). *The Autonomous Grid: Machine Learning and IoT for Utilities*. SAS Institute, Cary, NC.

Shaukat, N., Khan, B, Ali, S.M. et al. (2018). A survey on electric vehicle transportation within smart grid system. *Renewable and Sustainable Energy Reviews* 81: 1329–1349.

Speer, B., Miller, M., Schaffer, W. et al. (2015). *The Role of Smart Grids in Integrating Renewable Energy. NREL/TP-6A20-63919*. National Renewable Energy Laboratory; www.nrel.gov/publications/

Tan, K.M., Ramachandaramurhty, V.K., and Yong, J.Y. (2016). Integration of electric vehicles in smart grid: A review on vehicle to grid technologies and optimization techniques. *Renewable and Sustainable Energy Reviews* 53: 720–732.

8

Electric Power Management

Abstract

Electric power management is changing because of the growth of renewable power production. Demand for energy, of course, fluctuates over time and has some flexibility. For example, electric vehicle batteries can be charged when supply exceeds demand as part of an organized power management plan. On the other hand, solar- and wind-generated electricity are produced when the sun shines and wind blows, producing supply unconditional of demand. Electrical power demand management is needed because it is necessary to balance supply and demand at all times. Batteries have a critical value in this management as a place to store excess energy that is produced. The batteries are able to hold the excess energy and then supply it at peak power demand times. Time-of-use prices and real-time prices can be used to help regulate the balance between supply and demand. With modern electronics (smart meters and a smart grid) power companies have the capability to reduce demand at peak power times and increase demand when that is needed. To do this, mutually beneficial agreements are needed with customers.

8.1 Introduction

In the past, the electric power industry managed supply and demand by generating the amount of power (i.e. supply) that was needed to meet the demands of the system. With the growth of solar and wind generation, though, issues of supply and demand management have become more complex. One complicating factor is that solar and wind energy generation can create situations in which supply exceeds demand, a balance issue not typically addressed before. Another reason power management gets more complicated is because of a growing issue that arises when residents put solar panels on their roofs and those panels generate more energy than the household demands (e.g. during the day, while the residents are at work). The solution in this case is either for the power to flow into the electrical grid or to have household battery storage to save the electricity for later use.

This chapter will focus on balancing supply and demand within the context of a smart grid system, when solar and wind generation are present and they supply close to 100% of the power at some times. Energy storage in batteries and other systems are included both because stored energy can help to meet demand and because surplus power can flow into batteries for storage.

8.2 New Developments in Storage

Because solar and wind technologies do not have carbon emissions associated with electricity production, there has been great long-term interest and progress in the transition to these renewable systems. It is also important to understand, though, that the cost of electricity generated with solar and wind has declined significantly since 2010; this, along with decreases in battery costs, have created growing interest in energy storage technologies for balancing supply and demand of electric power. The economics of solar and wind generation are now such that a significant portion of new generation in 2017 and 2018 was via solar and wind. The vanguard locations for implementing these new systems, though, are places where the economics are most favorable.

For instance, Hawaiian electric companies have received very favorable proposals for seven solar generation plus battery storage systems to generate and deliver electricity on Hawaii, Maui, and Oahu. At the time of this writing, these are contracts that have been submitted to the Public Utilities Commission for review. The cost per kWh ranges from $0.08–$0.12 for systems, which include solar generation and battery storage (Colthorpe, 2019). Since the cost of electricity generation with imported fuel oil is about $0.15/kWh, the cost of the new system is projected to be very economically beneficial to the Hawaiian communities and to the reduction of greenhouse gas emissions as well. Batteries within such a system turn out to be superior to natural gas peaking plants because the batteries can be used to store excess power when supply exceeds demand.

It is also worth noting that, within this particular example, there are also solar panels on many homes and buildings in Hawaii. About 11% of total electricity demand in Hawaii is provided by solar. However, some of the utilities have needed to curtail excess rooftop solar energy that is available to feed into the grid because of grid management problems (Thurston, 2019). There are efforts now to upgrade the grids with smart meters and other improvements that will improve communication and grid management.

8.3 Demand Management

The adoption of smart meters and a smart grid opens up a range of possibilities for demand management. The development of demand management strategies will become increasingly important for the overall functioning of the electrical system. Smart meters and a smart grid can implement automated demand management that, for example, reduces power use in subtle ways when it is necessary to manage supply and demand. Some of this is currently being used in various locations, often based on the price of electricity where there is time-of-use pricing. People or organizations interested in saving money can use demand management to reduce their usage at peak power times (when demand is greater and time-of-use prices are high), sometimes by delaying activities until a time of lower prices when there is a need to increase demand. In particular, for locations where there is significant solar generation, there may be low prices for electricity in the late morning and in the middle of the day to encourage immediate and direct use of the solar-generated electricity. One good way to increase consumption during these time periods would be to plug in and charge electric vehicle batteries, which may be sitting in workplace parking lots.

Real-time prices may be used to manage supply and demand where customers have smart meters and there is good communication. Wholesale prices of electricity vary with supply and demand; these are the prices at which utilities buy and sell electricity from one another. Real-time prices for customers may be adjusted based on wholesale prices and/or the supply and demand of the utility (Taylor and Taylor, 2015).

Customers who allow their demand to be reduced by the utility are usually compensated for their participation in a direct load control program, where the utility reduces power flow to the customer during peak power times as needed. The air conditioner may run less or the electric water heater may receive less power, for example. One example of this approach is the program of Baltimore Gas and Electric. The Baltimore customers who allow their air conditioning to be reduced at peak demand times receive $1.25/kWh for each kWh that is not supplied (Taylor and Taylor, 2015).

Larger electricity consumers may participate in an interruptible load program, in which the utility may call and request that the load be reduced by an amount that has been agreed upon. There may be a time period of some minutes for the customer to reduce the load. In some cases, it may be possible for the utility to make the change remotely (Taylor and Taylor, 2015).

For programs such as those described above, the installation of smart meters (along with a smart grid; see Chapter 7) is a critical infrastructure that must be in place first. Smart meters have been very beneficial for advancing demand response programs because of the improved communication and better ability to educate customers with respect to the programs.

8.4 Future Supply and Demand Expectations

As the fraction of power generated via wind and solar energy increases, progress also needs to be made in finding new ways to balance supply and demand. It is already becoming clear that energy storage, such as with batteries, will be increasingly important. The capacities of batteries for energy storage can also be expected to increase in the future, just as they have in the past. There will continue to be a need, though, for on-demand electric power generation that can be produced when it is needed. The current expectation is that this need will probably be met by facilities such as natural gas power plants. These plants will be needed to provide power when the stored power is running low and more power is required to balance supply and demand.

There will also be a need for power sinks – locations and uses where excess power can be delivered to accomplish goals that are not time sensitive. These may be industrial processes that require large amounts of electricity. Since the cost of power is often a very important factor in terms of production economics for these industries, it makes a lot of sense to delay operations in order to take advantage of low energy costs. One example of this type of industry is the production of aluminum. It makes good economic sense to produce aluminum when and where the cost of power is low (Peck, 2015). Another example of this type of industry is the production of hydrogen.

Hydrogen production is important to society for several different applications (Baxter, 2018; IRENA, 2018). Hydrogen can be produced from electrolysis of water with oxygen as a co-product, and it can be stored in tanks. Small quantities of hydrogen can be added to natural gas lines and mixed with the methane gas. This blended gas can then be used as fuel to generate electricity when it is needed. The mixture can also be used for heating by burning the gas in a furnace or hot water heater. Hydrogen is also needed to make ammonia, which is used as a nitrogen fertilizer. Since the market for electricity varies with the seasons of the year, there is the potential to use excess electrical power to produce hydrogen via electrolysis of water during times of the day and times of the year when excess power is available.

Since both the efficiency of converting electricity to hydrogen using electrolysis is high and the efficiency of recovering energy using hydrogen-powered fuel cells is high, producing hydrogen when there is excess electricity is a good option. The applications for hydrogen-powered fuel cells include powering vehicles When these hydrogen fuel cells are used, the product of the reaction is water rather than pollutants. Thus, fuel cell-powered vehicles do not contribute to air pollution or carbon emissions.

The International Energy Agency has produced a new report on hydrogen which indicates that the future of hydrogen is very positive, with multiple uses as a versatile green product (IEA, 2019). The report points out that the costs of hydrogen from renewable electricity are expected to decrease as the costs of solar and wind energy decrease. Since the electrolysis of water to produce hydrogen can be used as a way to use electricity when the supply of electricity exceeds demand, there should be a lower price for the electricity that is used under these conditions and thus a lower cost of hydrogen production.

Storage of hydrogen in tanks is seen as an alternative to storing electricity in batteries. There is currently significant storage of natural gas in tanks because the demand for winter heating can be met by using gas that has been delivered and stored at an earlier time. There will be increased storage of hydrogen for transportation and other purposes. Hydrogen production in spring and fall, and storage of hydrogen for winter building heating and summer transportation, may have good value for society.

One of the issues with hydrogen is product safety. Some potential applications, such as a home burning of hydrogen in a furnace, may not be popular because of safety issues. Electric heating has already been developed and is used in many locations; it is considered to be a very safe way to heat a building.

References

Baxter, J. (2018). *Energy from Gas: Taking a Whole System Approach*. Institution of Mechanical Engineers, United Kingdom.

Colthorpe, A. (2019, January 7). Proposed solar-plus storage projects among Hawaii's lowest cost renewables ever. *Energy Storage News*; www.energy.storage.news/

IEA (2019). *The Future of Hydrogen*. International Energy Agency; www.iea.org/

IRENA (2018). *Hydrogen from Renewable Power: Technology Outlook for the Energy Transition*. International Renewable Energy Agency, Abu Dhabi.

Peck, M.J. (2015). *The World Aluminum Industry in a Changing Energy Era*. Routledge, New York.

Taylor, B. and Taylor, C. (2015). Demand response: Managing electric power peak load shortages with market mechanisms. *The Regulatory Assistance Project*; www.raponline.org/

Thurston, C.W. (2019, January 7). Rooftop solar curtailment to ease with refocused Hawaii energy contracts. *Clean Technica*; https://cleantechnica.com/

9

Off-Grid Power Management

Abstract

There are many people who would benefit from having electricity but who live too far from an electrical grid to access it. Recent developments can help such people. Solar-generated electricity and battery storage are now available at a very reasonable cost for those who want to have an electrical system in their home or business. The very simplest system of this type is a solar lantern that can be set in the sun each morning to charge the battery, and then used in the evening for lighting. Because of the low cost of solar-generated electricity now, there is an active transition around the world away from fuel oil or propane-powered generators and towards solar panels for generation of electricity. Lighting, cooking, communication, and refrigeration are among the most important reasons to have off-grid electricity. Storage batteries are an important part of the off-grid system.

9.1 Introduction: Electricity for All

There are more than 1 billion people who do not have electricity as of 2018 (Puliti, 2018). One of the Sustainable Development Goals is to have affordable, reliable, sustainable, modern energy for all people (IRENA, 2018). The World Bank has committed $1.3 billion to off-grid programs, and solar-generated electricity is one of the important technologies that is being advanced to bring electricity to all homes.

Why is it important that people have electricity? Lighting is one of the most basic needs that is included in this effort, although refrigeration and safe cooking are also important basic energy needs that are very beneficial to families. Increasingly, communication (e.g. by cellular telephone) is another high priority for many. All of these outcomes are enabled by access to energy. In places where there is an effort to bring electricity to those who are some distance from the grid, there has been good progress simply by ensuring access to supply chains and financing those communities (IRENA, 2018; Puliti, 2018).

A number of previous efforts for providing off-grid electrical service have been based on the premise of having an electricity generator powered by an engine that burns fuel oil or propane. Although this certainly has been a viable system, it also has a number of drawbacks: It creates an ongoing cost for the fuel, it is relatively susceptible to breaking down, and, of course, it is a source of pollution. Now that the price of solar panels has decreased, electricity generated by solar systems is less expensive and many newly installed systems include solar-generated electricity. Some off-grid systems now have solar panels, batteries, and a generator. When the battery storage is low but there is no solar energy, the generator is started and used to provide electrical power until the solar production is able to provide the desired amount of power.

Many solar home systems were installed as part of a large Renewable Energy Development Project in China, supported by the World Bank. The benefits included increased access to information, reduced use of candles and kerosene (reducing health and fire risks), increased use of basic appliances, and increased income for many of those who participated (van Gevelt, 2014). Van Gevelt (2014) reviews off-grid energy developments in many countries.

Policy actions and regulations have been beneficial in many countries. These actions are effective when they provide a strong foundation for market development, which encourages companies to provide services, supplies, and experienced employees to those who need and want off-grid power. These policy actions can also help with creating a stable need for education and training, in order to have qualified workers and customers who understand their system well enough to properly care for it and use it. In some cases, policies may provide incentives for people to participate in programs, and these need to be both understood and fairly implemented. Clarity and confidence in the plan that is being advanced are valuable, as are good standards and quality control to provide reliable products (IRENA, 2018).

9.2 Small Scale

The electrical grid is present in most cities and in some rural areas, but there are many parts of the world where there is no electrical grid. Over 100 million people have some electricity from off-grid systems that have been installed to serve their needs. Solar-generated electricity with battery storage can be used in these contexts at a number of scales. Application can range from mini-grids that provide energy for a household, group of houses, or even a small community, to one of the simplest developments: A solar lantern.

A solar lantern can be set out in the sun in the daytime to charge the battery and brought in for use in the evening. Solar lanterns are inexpensive and affordable for many of those who do not have access to the electrical grid (IRENA, 2018). This basic provision of evening lighting, however, can have transformative effects on people's safety, productivity, and well-being. Crucially, it is an application that does not require extensive training, expensive equipment, substantial maintenance, or even the existence of any electrical wiring in the home.

In Bangladesh, the Infrastructure Development Company (IDCOL) has had success with the installation of over 4 million solar home systems (SHS), which consist of a solar panel, battery, and LED light bulbs with associated wiring. An outlet to charge mobile phones is often included as well. This government-owned company has worked with the World Bank and other development organizations to provide the structure for the program. The program includes financing so that those who install an SHS are able to pay for it over a three-year period (Renner, 2017). Quality control has also been an important part of the program: Suppliers have to provide 20-year warranties for solar panels. Furthermore, there has been training to prepare those who work in the program to understand the systems and how to install them and provide services that are needed to make the program operate efficiently.

9.3 Mini-Grids

There are some geographical areas where a local mini-grid is appropriate for those who are presently living without electricity. The regular electrical grid may be prohibitively far away and there may be a growing number of people who want or need electricity, so a stand-alone mini-grid is clearly an option. Assuming there are not local resources (e.g. hydroelectric power options), solar and wind power are usually excellent options for these types of situations.

On the one hand, there are many benefits of linking the residents of the community together and building a mini-grid electrical system. The mini-grid can provide electricity reliably and with centralized expertise and quality control. On the other hand, there is also a need to address all of the policies and regulations that are needed to have appropriate rates and financing for a mini-grid. Maintenance and repair must be factored into the rates. There is also a need to have sufficient battery storage and/or generator power to provide for the power needs of the community. The members of the community and those in charge of the mini-grid need to have sufficient education to address these issues and operate the system successfully. Since there are capital costs of purchasing the solar panels, storage batteries, distribution lines, and all of the other equipment, financing must be found and included in the plan. There are resources to help with these considerations. For example, access to financing may be provided by the World Bank, the government of the country, or another source (IRENA, 2018). Without taking these considerations into account, though, mini-grids can suffer from instability or even collapse as a system.

9.4 Value of Electricity

The value of bringing electricity to populations that do not presently have it is significant. Employment opportunities are improved by bringing electricity to parts of the world that do not have it at present (Renner, 2017). There are jobs associated with the electrical systems that are installed, as well as other jobs because having electricity changes opportunities in the community. Residential living with electricity allows for purchases of appliances and other items that become useful because electricity is available. Work from home opportunities are enhanced by having electricity and better communication made available with electricity. The off-grid electrical systems can include those needed for business purposes, such as establishing a general store to sell food supplies and other locally needed products.

The efforts to provide lighting using solar energy have resulted in reduced use of kerosene for lamps and reduced use of candles. Electricity for cooking can reduce the use of wood or other carbon fuel-based fires. Both of these shifts have improved air quality in many homes and reduced greenhouse gas emissions, as well as reduced the risks of house fires. With the recent developments that have reduced the cost of solar-generated electricity and energy storage in batteries, many people in the world will be able to afford electric lighting and cooking.

The technology for off-grid electrical systems is well developed and available in a variety of publications (IRENA, 2018; Renner, 2017; Pirolli, 2016; van Eekhout, 2012). Pirolli (2016) lists four types of systems in the *Sustainable Energy Handbook*. These range from lighting only, to more complete electrical systems with appliances and back-up power provided by a generator.

With the present costs of solar-generated electricity and batteries, the potential to bring electricity to all homes in the world is financially possible. The simple solar lantern can be made available at a price of less than $50. A wired solar home system to provide lighting is more expensive, but within reach for many of those who do not have electricity at present. Where there is a will to accomplish these simple improvements for those living without electricity, it can be done.

References

IRENA (2018). *Off-grid renewable energy solutions: Global and regional status and trends*. IRENA, Abu Dhabi.

Pirolli, M. (2016). *Sustainable Energy Handbook*. Technical Assistance Facility, Western and Central Africa; https://europa.eu/capacity4dev/

Puliti, R. (2018). *Off-grid bringing power to millions*. The World Bank; www.worldbank.org/
Renner, M. (2017). *Rural renewable energy investments and their impact on employment*. International Labour Organization, Geneva, Switzerland; www.ilo.org/publns
van Eekhout, D. (2012). *Appropriate Modern Lighting Systems for Off-grid India* [master's thesis]. Utrecht University, The Netherlands.
van Gevelt, T. (2014). Off-grid energy provision in rural areas: A review of the academic literature; www.semantic scholar.org/

10

Policy

Abstract

Policies to reduce greenhouse gas emissions and improve air quality have been very important to the environmental progress that has been made. These policies vary across locations and also by the topics addressed, with larger locations (such as countries) often taking broader and more systemic changes, while smaller locations such as cities often take more pointed and innovative steps. Air quality has been improved in many locations because of these regulations, and efforts to improve air quality through regulations and policy incentives are continuing. This chapter reviews three exemplar locations and their policies. Norway has made great progress in reducing greenhouse gas emissions and improving air quality by adopting policies with incentives. China has used incentives and regulations to transition from motorcycles and scooters with internal combustion engines to electric bikes in its major cities. Southern California has made progress in a continuing effort to improve air quality in the Los Angeles metropolitan area using regulations and incentives.

10.1 Introduction

Most local, state, and federal governments have several policies that impact greenhouse gas emissions and air quality. For example, coal-burning power plants have air pollution controls because of regulations on emissions that impact air quality. Some of these policies also reflect the fact that electricity generation has been in transition to renewable sources, including wind and solar energy, for many years. In particular, some policies have created incentives to encourage wind and solar energy-generated electricity. There are nearly as many different policies as there are governments in the world. Nearly everywhere has approved policies that are now being followed, some of which are more extensive and progressive than others.

The US Clean Air Act is the focus of Chapter 7 in the book *Choked* by Beth Gardiner (2019). The history and politics of legislative events associated with the Act are described, as well as the great success in reducing early deaths associated with air pollution by more than 150,000 every year. The ratio of benefits to costs is estimated to be 30 times more benefits than costs.

The policy developments associated with the Clean Air Act have improved the quality of life for many people. The health benefits of air pollution regulations are discussed in Chapter 3 and in two papers that are publicly available (Erickson, 2017; Erickson et al., 2017).

Examples of the policies that are of interest for this topic include those related to the generation of electricity, those related to transportation, those related to carbon emissions, and those related to air quality. The Paris Agreement on Climate Change has resulted in policy actions by a number of governments, and it is one of the reasons for many of the policy actions that are described in this chapter.

Among the reasons why policies differ from place to place is that energy policy may need to be different for developing countries compared to more developed countries. The best pathways to reduce carbon emissions in any given region should consider the costs and benefits of implementing renewable energy technologies, as well as the ability of the citizens to pay for those advances. One of the truisms of economic development is that carbon emissions will increase with growth in income per capita and with population size. This growth in emissions with economic development, though, can be slowed by transitioning to renewable energy and electric vehicles alongside that development. In fact, adding renewable energy generally has a positive impact on economic development (Valadkhani et al., 2019), in addition to the reduced emissions that are beneficial for health.

Because of the impact of climate change on the quality of life and health, many developed countries have implemented carbon taxes and emissions trading schemes. The European Union, for instance, has been very successful in reducing greenhouse gas emissions by almost 20% over the period from 1990 to 2012 (Villoria-Saez et al., 2016). A large proportion of that EU success has been due to carbon taxes and emission trading schemes, implemented to provide incentives to reduce carbon emissions.

British Columbia (Canada) introduced a revenue-neutral carbon tax in 2008. This tax covers about three-quarters of all carbon emissions in the province. The tax has resulted in a 5–15% reduction in greenhouse gas emissions. Although the effect on the economy has been very small, there is good public support for the tax (Murray and Rivers, 2015).

10.2 Electricity Generation Policy

Coal combustion is still extensively used in many countries to generate electricity. Because it produces more carbon emissions than other processes to generate electricity, there are many existing policies that are related to efforts to reduce greenhouse gas emissions from coal use and improve air quality. There are restrictions on particulate, sulfur, and mercury emissions that require expensive emission controls on coal plants. These alone are helping to shift the economics in favor of natural gas, wind, and solar energy for power generation. More recent policy efforts in many countries have tried to help guide this transition to other methods for generating electricity.

A number of countries plan to phase out coal-generated electricity by 2030 (PPCA, 2019). In 2019, the Powering Past Coal Alliance has 80 members that are working cooperatively towards the goal of phasing out unabated coal power generation of electricity by 2030. Thirty of those members are national governments. Part of the motivation for these policies to phase out coal combustion is in order to meet the goals of the Paris Climate Agreement, and another part of the motivation is based on the desire to improve air quality. The Powering Past Coal Alliance declaration states that "more than 800,000 people die each year around the world from the pollution generated by burning coal".

The transition away from coal has started. Electricity production from coal peaked in 2014, and many coal-fired generating plants are now running at lower production levels because electricity from wind and solar are less expensive operationally. That is, it is now more cost effective to use the electricity generated from wind and solar, and to maintain the coal plant at a reduced power level for supply and demand management. The capacity to generate electricity from coal has grown since 2014, even as production started falling, due to new plants and new technologies. Even coal power capacity may have peaked as of 2017, because many old coal-fired power plants are being retired from service. The number of new coal-fired generating systems in 2017 was very small (Carbon Brief, 2018). In 2016, the global estimate was that 52.5% of coal generating capacity was being used to generate electricity (Carbon Brief, 2018). Belgium has stopped using coal to generate electricity, and ten other European countries have pledged to phase out coal-fired generation by 2030 (PPCA, 2019; Carbon Brief, 2018).

Of note is the fact that natural gas is included here as an alternative to coal-fired generation of electricity. Indeed, there are several reasons to favor natural gas in this scenario. Combustion of natural gas to generate electricity reduces carbon emissions significantly compared to coal, and natural gas is a cleaner process with respect to air pollution emissions. Natural gas prices have also fallen dramatically in the past couple of decades, making it more affordable.

When comparing all energy production methods, however, wind and solar energy generate electricity without any combustion emissions. Wind and solar energy also generate electricity without an expensive infrastructure for extracting, refining, transporting, and burning a fuel. Thus, there have been significant incentives to encourage the use of renewable energy to generate electricity both in many countries and in many parts of the

United States. Research funding to help aid this goal has been provided in many countries, and this has resulted in significant progress in reducing the cost of wind and solar power generation. Other research has also contributed to progress to reduce emissions associated with coal-fired power plants.

In the United States, 29 states have established renewable portfolio standards, in which a requirement is established for a percentage of electricity to be generated through renewable processes (Rountree, 2019). These standards are considered to be the most important policy actions by states to advance the generation of electricity from solar and wind energy and to reduce greenhouse gas emissions. These policies have helped to create markets for wind- and solar-generated electricity and to develop employment opportunities as well. As a result of advances in renewable technologies, prices for electricity are lower compared to where they would be without these developments (Rountree, 2019).

Because of research and development progress, wind and solar energy have become much more competitive for generating electricity, and global progress is being made in adding new generating capacity. In fact, some of these incentives to add new solar and wind generating capacity are being reduced in the United States, China, and other countries because the economics are changing as new developments are implemented.

Relaxing or removing these incentivizing policies is economically feasible at this stage, but that does not mean that such changes are wise. We must continue to encourage electricity generation from renewable energy sources for the reduction of greenhouse gas emissions and the goal to improve air quality. For example, the transition to electric vehicles has begun in many respects, but it can have a greater impact on improving air quality if the electricity that electric vehicles use is generated from renewable energy. Policy developments to replace combustion of coal and natural gas with zero-carbon energy generation sources continue to have value with respect to reducing emissions.

More important and needed changes in policies are those that address the secondary consequences of the transition to wind and solar energy. The advances in adding wind- and solar-generated electricity to the grid have resulted in a need to transition to a smart grid and manage that grid differently. There are changes in policy to address these new sources of power with incentives to install new smart grid technology and encourage its use. New rate structures have also been introduced and communication with users is being improved as part of this new infrastructure.

The developments in solar energy technology and affordability have also resulted in many individuals installing solar panels on their homes to generate electricity for their own use (Geng et al., 2017). Because of variations in supply and demand, sometimes the resident is using electricity from the solar panels only, sometimes the resident is using a mix of solar power and power from the grid, and sometimes excess electricity can flow from the solar panels to the grid. Policies have been developed in many parts of the world to manage these distributed power generation arrangements. In some locations, net metering has been implemented, and distributed electricity generated by residents and businesses may flow into the grid and be valued the same as that which flows from the utility.

10.3 Transportation

Many policies have been enacted to address transportation because of the significant effects of combustion emissions from transportation on air quality and because of the desire to reduce greenhouse gas emissions. These include policies motivated by health impacts associated with diesel emissions, which are a major concern in many countries. Incentives to purchase battery electric vehicles (BEVs) and plug-in hybrid electric vehicles (PHEVs) are currently common in many countries.

Norway, for example, has made great progress in its efforts to encourage the purchase and use of electric vehicles (EVs). The country has set a goal to have all new cars with zero emissions by 2025 and it is on track to reach this goal, with more than half of new car sales being plug-in models in 2017. Norway has implemented an array of incentives to work towards this goal. EVs have reduced purchase taxes and are exempt from the 25% value added tax. Road taxes, road tolls, ferry prices, and parking costs are also less for EVs. These financial incentives have resulted in Norway being the third largest market for EVs in the world in 2017 (Jones, 2018).

In a very different part of the world, air pollution in Los Angeles has been a major health concern for more than 50 years. This has led to an ongoing effort to improve air quality through regulations and other policies within the LA area and in California overall. California has had significant success with incentives to encourage EV sales. About half of the EV sales in the United States have been in California, where there are significant policy goals and incentives to encourage EV sales and use. The policy incentives are in place because of air pollution and the goal to improve air quality and health. Secondarily, though, California is also attempting to reduce greenhouse gas emissions.

There is a clean vehicle rebate in California when a new EV is purchased; the amount in 2019 is $2500 for all EVs, such as the Chevy Bolt, and $1500 for many PHEVs (DriveClean, 2019). There is also a Clean Air Vehicle decal, which allows drivers of an EV special lane access. Air pollution regulations in the United States more broadly have been impacted by the development of regulations and policies in California (designed in particular to improve air quality in Southern California). Thus, there is now also a federal tax credit of up to $7500 for the purchasing of EVs in all states. The value of the tax credit depends on the size of the battery and how many EVs have been sold by the manufacturer.

Currently, there are efforts in the California Legislature to pass legislation to have all new cars sold in California be EVs by 2040. The goal of AB 40 for 2040, introduced by Phil Ting, is to have the California Air Resources Board develop a plan by January 1, 2021 that would enable California to achieve this goal by 2040 (Morris, 2018). It will be interesting to see if this policy also expands in some form to the rest of the United States.

A number of cities and countries have developed goals and/or plans to reduce greenhouse gas emissions and improve air quality by encouraging the electrification of transportation. Eighteen countries are included in a table of targets to phase out passenger vehicles with only internal combustion engines (ICEs) (Burch and Gilchrist, 2018). The earliest date for this is in Austria, where the plan is to not sell new ICE vehicles after 2020. India has a goal of no new ICE cars sold after 2030. The goal for Norway is to transition to all new car sales being EVs by 2025. The United Kingdom and France have a goal of ending sales of new cars with only an ICE by 2040. China plans to end production and sales of ICE vehicles by 2040.

A different, but not mutually exclusive, approach from the above policies encouraging EV adoption is to have regulations that restrict ICEs in areas where efforts are being made to reduce greenhouse gas emissions and improve air quality. In China, for instance, motorcycles are restricted in large cities and this has resulted in electric bicycles being used in their place. A number of cities have areas where vehicles with ICEs are restricted. Others have areas where a permit must be purchased daily, and the price of the permit depends on the emissions of the vehicle (Erickson, 2017; Erickson et al., 2017)

In large cities where air quality is a concern, actions can be taken to transition to electric buses, electric taxi services, electric delivery vehicles, and electric garbage trucks (see Chapter 4). EV charging infrastructure can be installed to encourage commuters to have EVs. Policies to encourage progress in these efforts have been developed in many cities (Erickson et al., 2017).

There is a list of 25 cities that have regulations, plans, or goals to reduce air pollution due to vehicles with an ICE in Burch and Gilchrist (2018). Delhi has a deregistration program for diesel cars that are more than ten years old. These old cars cannot be driven in the city. Diesel cars are banned in some cities (Tokyo, Paris). A number of cities have signed the C40 Fossil-Fuel-Free Streets Declaration to have electric buses by 2025 and no ICE vehicles by 2030 (Burch and Gilchrist, 2018; Lipton et al., 2018). The Zero Emissions Vehicle Network is a cooperative effort for cities to work together on zero-emission vehicle strategies for whole cities or areas of cities. Charging infrastructure for EVs, incentives, and fleets are included, with electric buses, taxis, and municipal vehicles included in the fleet effort (C40 Cities, 2018).

Lastly, China has some new policies that became effective in 2019. All major auto companies doing business in China are required to meet minimum requirements for producing electric, plug-in hybrid, or fuel cell vehicles (Campbell and Ying, 2019). The target for 2019 is 10%, which rises to 12% in 2020, and the goal is to increase to 7 million in 2025 (about 20% of new car sales being plug-in hybrids, EVs, or fuel cell models). To encourage the purchasing of cars without emissions, the largest cities, such as Beijing, are restricting the issuance of new license plates and charging as much as $14,000 for some vehicles with ICEs, as compared to zero charge for an EV (Campbell and Ying, 2019). These new policies in China are having worldwide impacts on the auto industry because the market for new vehicle sales is very large in China. More than 1 million plug-in hybrid and EVs were sold in China in 2018, about 4.2% of new auto sales; in January 2019, plug-in vehicle sales were 4.8% of new auto sales (Davis, 2019; Pontes, 2019).

10.4 City Policies

Nations and large states tend to work with large-scale and incremental policies, which are important and effective for such large and diverse entities. Individual cities can be different in a number of ways. Cities often face more specific issues, lending themselves to more pointed possible solutions. Having a relatively smaller and more unified population, there is sometimes also an ability for cities to be more innovative and progressive in their policies to address their concerns.

The core issues for cities, though, remain very much the same. Many cities have developed policies to reduce greenhouse gas emissions and to improve air quality. Some cities are using the C40 Cities Climate Action Planning Framework (Lipton et al., 2018) to develop their climate action plan. They are transitioning to renewable energy for their electricity, developing an EV charging infrastructure, purchasing electric buses, encouraging the transition to electric taxis, improving building energy efficiency, adding bicycle paths, and increasing pedestrian options for people to walk to their destinations.

There is an air quality initiative whereby cities are working cooperatively to improve air quality by measuring air quality with inexpensive monitoring instruments, engaging citizens in actions to address air pollution, and developing new policies and enforcing existing regulations that improve air quality and health (C40 Cities, 2018). One of the values of climate action planning is the public commitment to achieving the goals of the plan. All residents of a city can participate and reduce their carbon emissions. Tracking progress and communicating the results of sustainability efforts provides encouragement to all those who are helping with the efforts.

Copenhagen, Denmark has a plan to achieve carbon neutrality by 2025. A 42% reduction in carbon emissions has already been achieved compared to 2005 levels, with the majority of the 66 initiatives in the first plan having been implemented. Streetlamps, for example, have been replaced with LED lights. There is a periodic review and revision of this plan, and total emissions are reviewed and benchmarked for each year. Electric buses have been purchased and put into use. Efforts are being made to improve bicycle lanes to encourage more bike transportation. In many Copenhagen locations, there are lanes for walking, other lanes for biking, and other lanes for motor vehicles (Lipton et al., 2018).

Paris adopted a climate action plan in 2007 to reduce greenhouse gas emissions by 25% by 2020, and to reduce emissions by 75% by 2050 relative to 2004. Work was started in 2016 on a new climate action plan that includes a goal of zero carbon emissions by 2050 in central Paris and an 80% reduction of the carbon footprint of the greater Paris area by 2050. All energy is to come from renewable sources by 2050 (Lipton et al., 2018).

New York City has a plan to align itself with the Paris Agreement. This plan includes 30 near-term actions, to be started by 2020, with the goal of reducing carbon emissions. Carbon emissions savings potential and the potential benefits, costs, and feasibility were considered in developing this plan. Growth, equity, sustainability, and resilience were also city goals that were used, along with 13 potential benefits, such as quality jobs and workforce development in prioritizing options (Lipton et al., 2018).

Oslo, Norway has a plan to reduce greenhouse gas emissions by 36% by 2020, and 95% by 2030. Transportation, energy and buildings, and resources (water and waste) are all included in the plan. Transportation is the most important element for the Oslo plan because 60% of the city's carbon emissions are associated with transit, and there is a good understanding of how to reduce emissions. Energy and buildings account for 20% of Oslo's carbon emissions, and here there is an effort to reduce the use of energy from fossil fuels by replacing it with electricity from hydro power (Lipton et al., 2018).

Stockholm, Sweden has a goal to achieve net zero greenhouse gas emissions by 2040. The target for 2020 is to reach the equivalent of 2.2 tons of carbon dioxide emissions per resident, per year. Reduced use of energy in buildings and the electrification of transportation are important elements of this overall goal. Differentiated taxes may be introduced to encourage use of EVs (Lipton et al., 2018).

Policies across all cities are important with respect to energy use in buildings, because heating, air conditioning, lighting, electrical appliances, and communication devices all use energy. The most important policy issue on this front is building energy codes, because insulation in high-quality buildings with good energy performance requirements helps to reduce energy use in buildings. Generally speaking, energy efficiency requirements for appliances, motors, and electronic equipment have the potential to significantly reduce energy consumption. The carbon emissions associated with buildings in the residential sector are about 17% of global emissions.

All buildings account for about 40% of energy consumption. Public education related to building energy conservation and efficiency has value as well (Nejat et al., 2015). More than 32 countries have building energy codes, so further progress in establishing building energy codes has the potential to reduce energy use in buildings by 50% (Nejat et al., 2015). Finally, in order to dramatically reduce greenhouse gas emissions associated with buildings, the electricity supplied needs to be from zero-carbon sources such as solar energy.

10.5 Conclusions

Policies enacted to reduce greenhouse gas emissions and improve air quality have been very valuable in many cities, states, and countries. Progress has been impressive but uneven around the world. China has been making great progress to improve urban air quality through policy implementation. California has improved air quality by implementing regulations that have reduced emissions and by using incentives to encourage the transition to EVs. Norway has used policies to increase the number of EVs that are purchased and used. Policies that reduce carbon emissions by transitioning to renewable energy for generation of electricity and EVs have health benefits because of better air quality.

References

Burch, I. and Gilchrist, J. (2018). *Survey of Global Activity to Phase Out Internal Combustion Vehicles*. Center for Climate Protection [September 2018 revision]; www.climateprotection.org

C40 Cities (2018). *Zero Emission Vehicles*. C40 Cities, London, UK; www.c40.org/; see also Air Quality.

Campbell, M. and Ying, T. (2019, January 6). Electric vehicles. *Bloomberg Business Week 2019 The Year Ahead*, 36–37.

Carbon Brief (2018, June 5). Mapped: The world's coal power plants; www.carbonbrief.org/

Davis, C. (2019, January 15). China electric car sales soar to almost 160,000 in December. *Insideevs*; https://insideevs.com/

DriveClean (2019). *DriveClean.ca.gov*; www.driveclean.ca.gov/

Erickson, L.E. (2017). Reducing greenhouse gas emissions and improving air quality: Two global challenges. *Environmental Progress and Sustainable Energy* 36: 982–988.

Erickson, L.E., Griswold, W., Maghirang, R.G., and Urbaszewski, B.P. (2017). Air quality, health, and community action. *Journal of Environmental Protection* 8: 1057–1074.

Gardiner, B. (2019). *Choked: Life and Breath in the Age of Air Pollution*. University of Chicago Press, Chicago.

Geng, Y., Chen, W., Liu, Z. et al. (2017). A bibliometric review: Energy consumption and greenhouse gas emissions in the residential sector. *Journal of Cleaner Production* 159: 301–316.

Jones, H. (2018, July 2). What's put the spark in Norway's electric car revolution? *The Guardian*; www.theguardian.com/

Lipton, J., Miclea, C., Fernandez, I. et al. (2018). *Cities Leading the Way*. C40 Cities Climate Leadership Group, London, UK; www.c40.org/

Morris, J.D. (2018, December 3). California lawmaker tries again to gradually ban gas cars. *San Francisco Chronicle*; www.sfchronicle.com/

Murray, B. and Rivers, N. (2015). British Columbia's revenue-neutral carbon tax: A review of the latest "grand experiment" in environmental policy. *Energy Policy* 86: 674–683.

Nejat, P., JomehZadeh, F., Taheri, M.M. et al. (2015). A global review of energy consumption, CO2 emissions, and policy in the residential sector (with an overview of the top ten CO2 emitting countries). *Renewable and Sustainable Energy Reviews* 43: 843–862.

PPCA (2019). *Powering Past Coal Alliance*; poweringpastcoal.org/

Pontes, J. (2019, February 24). China electric vehicle sales jump 175%, up to 4.8% of auto market in January! *Clean Technica*; https://cleantechnica.com/

Rountree, V. (2019). Nevada's experience with the Renewable Portfolio Standard. *Energy Policy* 129: 279–291.

Valadkhani, A., Nguyen, J., and Bowden, M. (2019). Pathways to reduce CO2 emissions as countries proceed though stages of economic development. *Energy Policy* 129: 268–278.

Villoria-Saez, P., Tam, V.W.Y., Merino, M.D.R. et al. (2016). Effectiveness of greenhouse gas emissions trading schemes implementation: A review on legislations. *Journal of Cleaner Production* 127 49–58.

11

Economics

Abstract

Economic considerations impact many decisions relevant to reducing greenhouse gas emissions and improving air quality. Research and development to improve the economics of electricity from solar and wind energy has enabled these technologies to become financially competitive by a variety of standards. For example, as battery prices have fallen, electric vehicles (EVs) have become more competitive with gasoline-powered vehicles, a relatively simple comparison. Although the larger transition to reduce greenhouse gas emissions is in progress, many consumers have not yet made the transition to electric bikes and EVs. Because of the costs of climate change and poor air quality, there are presently good economic reasons to continue the transition to renewable energy for electric power generation and to EVs for transportation. In large cities, there are also social and economic benefits associated with the electrification of public transportation. The benefits of improving air quality far exceed the costs.

11.1 Introduction

Economics is very often a consideration in decisions that are made about greenhouse gas emissions and air quality, and so it is a very important topic with respect to these issues. However, these issues can become thorny when viewed through an economic lens, for several reasons. One reason is that the scale of economic decisions regarding greenhouse gas emissions and air quality so often involve immensely large numbers: Millions, billions, and trillions of units (e.g. monetary costs and benefits, emissions, and subsidies). Another reason is that these economic decisions often deal with phenomena happening over large timescales: Years, decades, and even centuries. Finally, there are thorny aspects in terms of the different types of costs and benefits; some of the costs and benefits are personal and immediate to individuals (or to organizations), whereas other costs and benefits are general, long term, and spread across the population (see Chapter 15).

This chapter gives an overview of the economics underlying the reduction of greenhouse gas emissions and the improvement of air quality. Some big numbers and long time frames are inevitable, but we try to make the economics more understandable where possible. This chapter takes the position that all the costs and benefits – across individuals, communities, nations, and the globe – should be considered when assessing the economics of these issues.

To start with, the transition itself, from coal-fired power plants to renewable electric power, is economically a very large process. There are thousands of coal-fired power plants, and as they age they tend to gradually become more expensive to maintain and run. Some existing coal plants are over 50 years old now. At the same time, the prices for electricity generated by wind and solar energy have decreased, and continue to decrease. Thus, the economics associated with the transition from coal to wind and solar are much better in 2019 compared to just 20 years ago.

The transportation transition from combustion power to electric vehicles (EVs) with batteries similarly depends on economics and advances in science and technology. Research to develop EVs with better and less expensive batteries has been ongoing for many years, with consistent advances during that time. Significant progress has been made and battery prices have decreased to less than $200/kWh as of 2018. If customers purchase EVs because they are considered to be the most economical choice, this will help to advance the transition to EVs. This works hand in hand with the other important advances in wind and solar energy, which can be used to generate the electricity for those EVs (see Chapter 4). As the prices decrease for electricity generated with renewable technologies, and the purchase prices of EVs decrease as well, these will reciprocally influence each other. Once carbon-free electricity has a cost advantage over alternatives, it becomes easier to expand the amount that is generated.

The needed economic activity associated with transitioning from combustion processes to renewable energy for both transportation and the generation of electricity is of the order of trillions of dollars per year globally. The importance of reducing combustion emissions by more than 80% by 2050 is appreciated and recommended by many (Erickson et al., 2017), and this can be accomplished by electrifying transportation and generating all electricity with renewable energy (Williams et al., 2012; Erickson, 2017). Since the global social cost of poor air quality in large cities is of the order of $3 trillion/year, there is a significant incentive to take action and move forward with the transition away from combustion processes (Erickson, 2017; Scovronick et al., 2019). Scovronick et al. have added the benefits of better air quality to economic cost/benefit models that address greenhouse gas reductions, and they have found that actions that reduce emissions and improve air quality have immediate benefits for society in many parts of the world, especially in large cities where air quality is poor. The global health benefits of reducing greenhouse gas emissions along an optimal path are of the order of $1 trillion/year when health is included in the model (Scovronick et al., 2019). The largest health benefits are in India and China.

What about the alternative? If the world decided to do nothing to address climate change, including the contributions from greenhouse gas emissions and air pollution, this would progressively exceed the costs of taking action. When exactly these costs would exceed immediate benefits is debated, but there is no debate that it will happen eventually. Overall, the negative economic impact on production of a changing climate has been estimated to be about a 23% reduction in economic productivity by 2100 (Burke et al., 2015).

11.2 Costs and Benefits Related to Reducing Greenhouse Gas Emissions

One of the economic challenges is to estimate the costs and benefits associated with potential changes that can be made to reduce greenhouse gas emissions and improve air quality. An important early study on the costs and benefits of taking action to reduce carbon emissions was conducted by Stern (2007), which made a strong case that the benefits of reducing emissions far exceed the costs.

First, let's look at the costs. The social cost of carbon, which is the cost to society associated with a unit of greenhouse gas emissions, is considered in the report and related to the cost to reduce emissions. This value is about $46/ton of carbon dioxide (Gillingham and Stock, 2018). There is value in bringing the social cost of carbon into economic analysis. Some of the impacts of carbon emissions include melting glaciers, flooding, declining agricultural yields, heat stress, rising sea levels, displaced populations, more wildfires, and ecosystem impacts (Stern, 2007). When the quality of life in developing countries is reduced because of climate change, there may be significant migration, which has effects in other locations. The Stern report indicates that there may be a reduction of 5–20% in global economic activity because of greenhouse gas emissions if no actions are taken to reduce emissions. This compares with the 23% reported more recently (Burke et al., 2015). On these facts alone, one might conclude that taking action to reduce greenhouse gas emissions is both feasible and highly desirable.

There are also significant social and economic challenges associated with reducing emissions globally. To achieve reductions in greenhouse gas emissions, the Stern report recommends systemic revisions to carbon emissions prices, technology policies, and better opportunities for behavioral change. If carbon emissions prices can be global, this would make such recommendations practical and feasible. Using the $46/ton mentioned previously, and the global emissions of 33.1 billion tons of carbon dioxide in 2018, a carbon tax would generate $1.52 trillion/year, which could be used in positive ways to advance efforts to transition to renewable energy and electric vehicles, improve quality of life, and adapt to climate change impacts (IEA, 2019). Public education related to technology developments and positive behavioral changes that are beneficial may also be included.

It is true that these transition efforts would cost money. Minimizing the cost of mitigation and maximizing the benefits of actions that are beneficial have been addressed (Gillingham and Stock, 2018). One way to describe these cost/benefit relationships is in terms of how much it would cost to eliminate one ton of carbon dioxide emissions which would otherwise occur. The concept is that these cost estimates incorporate the replacement of a fully paid for coal plant with a new plant (Gillingham and Stock, 2018). Cost estimates to reduce one ton of carbon dioxide emissions include $25 for onshore wind, $27 for natural gas combined cycle, and $29 for utility-scale solar photovoltaics.

In all of the above estimates, the dynamics of costs are impacted by research and development and by the evolution of manufacturing operations. An analysis of solar-generated electricity with solar panels shows that there were significant benefits to having incentives to encourage the purchase of solar panels, and from supporting research to develop more efficient solar photovoltaic technologies. As a result, the prices of solar panels have decreased more than 70% (Gillingham and Stock, 2018). Therefore, the $29/ton reported earlier is almost certainly lower now because of the very substantial progress in cost reduction that has been accomplished.

Economic issues related to EVs are also included in Gillingham and Stock (2018), but the numbers reported may similarly be out of date now. The dynamic decreases in prices due to battery developments have particularly impacted progress on EV innovations and sales. The number of EV sales and operations has changed because of competition, battery developments, and policies to encourage EV purchases. Some manufacturers are transitioning to new EV models because they do not want to be left behind, and they want to have the mix of models needed to sell in China. Economic incentives in many other countries have been successful in increasing the sales of EVs (see Chapter 4).

There are also other methods to reduce carbon emissions, and some of these are very inexpensive per ton of carbon dioxide. Reforestation of land has a range of values from $1–10. The Clean Power Plan has an estimated value of $11. Renewable portfolio standards have a range from $0–190. This large range reflects the fact that the economics of wind and solar energy depend on location (Gillingham and Stock, 2018).

11.3 Costs and Benefits Associated with Improving Air Quality

The costs and benefits associated with improving air quality overlap substantially with those associated with reducing greenhouse gas emissions (and climate change generally), but there are some distinct issues for air quality. The specific air quality issues tend to involve more direct and immediate consequences for health, including everything from general quality of life, to illness and time lost from work, to effects on mortality.

The most significant benefits from improving air quality can usually be enjoyed in urban communities. This has to do with two factors: Cities tend to have the highest rates of poor air quality (due to the amount of pollution being emitted) and cities have the largest populations in a compact area (high population density). Thus, the transition to combustion-free processes, both to generate electricity and for transportation, has huge potential to improve life in large cities. The progress in reducing the costs of electricity from wind and solar energy, along with the reductions in battery costs, are contributing to a reduction in the costs of improving air quality. In summary, this again is a moving target in terms of economic analysis. Nevertheless, it is clear that the social value of living in an air environment that meets all of the regulatory standards is very great.

If all large cities met the air quality guidelines of the World Health Organization (WHO), the annual social cost of air pollution would decrease by more than $2 trillion because many large cities have air quality with contaminant concentrations that far exceed the values recommended by the WHO (WHO, 2018). The analysis of costs and benefits associated with the US 1990 Clean Air Act Amendments shows that the estimated benefits are of the order of $1.9 trillion per year, and the costs are of the order of $65 billion per year in 2020 (US EPA, 2011). To put those numbers in context relative to each other, the annual benefits are roughly 30 times greater than the annual costs. Among the largest costs in the above figure are those associated with fuels and pollution control devices for on-road vehicles (about $28 billion). Costs such as these will decrease as EVs become more common.

Efforts to improve air quality have been a high priority in California for more than 50 years. The recent emphasis on the electrification of transportation has helped, along with many other steps, to improve air quality. In Southern California, the ozone concentration has decreased over time from concentrations greater than 300 ppb to values that are less than half of that (Parrish et al., 2016). The particulate matter (PM2.5) concentrations have been decreasing by about 6.2% per year. The latest data show that air quality in the Los Angeles metro area still does not meet air quality standards on many occasions (albeit on fewer occasions than in the past). On more than 100 days per year, the ozone concentration exceeds the recommended value (ALA, 2018). Progress has been better in terms of compliance with the regulations for PM2.5. Even as only a partial success, though, the significant health benefits of improving air quality have far exceeded the costs in Los Angeles (Parrish et al., 2016).

In Los Angeles, as with elsewhere, there is a continuing need to reduce emissions, and efforts to do this are moving forward. This is not just a part of the efforts to reduce greenhouse gas emissions and address climate change, but also because the benefits of improved air quality on people's quality of life is justified on its own. Economic analyses of these effects from better air quality can be difficult to conduct because many of the benefits are difficult to quantify (e.g. quality of life, fewer/less severe illnesses, fewer childhood health issues, fewer premature deaths). When these effects are estimated, though, they are substantial benefits relative to the costs.

11.4 Government Economic Action

A number of government actions have impacts on greenhouse gas emissions and air quality (see Chapter 10). Inexpensive energy is beneficial because it reduces the costs of manufacturing products, heating, air conditioning, and transportation. Import and export prices are impacted by the cost of energy, and the balance of trade is impacted by energy subsidies that governments implement. Because inexpensive energy is tied to so many other things – things that governments often are quite interested in promoting – energy production costs can be a target for government actions. Thus, it is sometimes necessary to look at not only the direct costs of energy generation, including environmental impacts from that process, but also the government taxes and incentives that may be in place to create a lower energy cost experience than the actual generation and impacts (i.e. true costs) would otherwise create.

The International Monetary Fund has published a working paper in which the authors have attempted to estimate the subsidies associated with fossil fuels (Coady et al., 2019). Global subsidies were estimated to be $4.7 trillion in 2015 and $5.2 trillion in 2017 (6.3% and 6.5% of global GDP in 2015 and 2017 respectively). The estimated subsidies for specific countries in 2015 were: $1.4 trillion for China, $649 billion for the USA, $551 billion for Russia, $289 billion for the EU, and $209 billion for India. Coal is the largest source of subsidies at 44%, and petroleum is 41%, followed by natural gas at 10%. These subsidies distort the true costs of energy production, and they alone provide a strong economic case for moving away from those energy production means. Over and above all of that are all the good economic reasons to reduce greenhouse gas emissions by transitioning to renewable energy to generate electricity and zero-emission transportation options.

References

ALA (2018). *State of the Air 2018*, American Lung Association.

Burke, M., Hsiang, S.M., Miguel, E. (2015). Global nonlinear effect of temperature on economic production. *Nature* 527: 235–239.

Coady, D., Parry, I., Le, N.PP., and Shang, B. (2019, May). Global fossil fuel subsidies remain large: An update based on country-level estimates. International Monetary Fund; www.imf.org/

Erickson, L.E. (2017). Reducing greenhouse gas emissions and improving air quality: Two global challenges. *Environmental Progress and Sustainable Energy* 36: 982–988.

Erickson, L.E., Robinson, J., Brase, G., and Cutsor, J. (2017). *Solar Powered Charging Infrastructure for Electric Vehicles: A Sustainable Development*. CRC Press, Boca Raton, Florida.

Gillingham, K. and Stock, J.H. (2018). The cost of reducing greenhouse gas emissions. *Journal of Economic Perspectives* 32: 4, 53–72.

IEA (2019). *Global Energy and CO2 Status Report*. International Energy Agency.

Parrish, D.D., Xu, J., Croes, B., and Shao, M. (2016). Air quality improvement in Los Angeles: Perspectives for developing cities. *Frontiers in Environmental Science and Engineering* 10 (5): 11, 1–13.

Scovronick, N., Budolfson, M., Dennig, F. et al. (2019). The impact of human health co-benefits on evaluations of global climate policy. *Nature Communications* 10: 2095; https://doi.org/10.1038/s41467-019-09499-x

Stern, N. (2007). *The Economics of Climate Change*. Cambridge University Press, Cambridge, UK.

US EPA (2011). *The Benefits and Costs of the Clean Air Act from 1990 to 2020*. US Environmental Protection Agency, Washington, DC.

WHO (2018). *Ambient (Outdoor) Air Quality and Health*. World Health Organization, Geneva, Switzerland.

Williams, J.H., DeBenedictis, A., Ghanadan, R. et al. (2012). The technology path to deep greenhouse gas emission cuts by 2050: The pivotal role of electricity. *Science* 335: 53–59.

12

Agriculture

Abstract

There are significant issues associated with agriculture and its relationships with greenhouse gas emissions and air quality. Just as in other areas, energy is needed for farming, and transitioning to renewable energy in agriculture is an important goal. Because farms are large in area, the electrical grid has higher costs of transmission and distribution in rural parts of the world. There are many places where there is no electrical grid serving farm populations. The decreases in the costs of wind- and solar-generated electricity and the progress in battery development are very beneficial for agriculture and off-grid electrical systems. Sustainable development in agriculture is a high priority and has great value for society. Nutritious food is one of the most important needs for good health and well-being. Food security for all people is one of the Sustainable Development Goals. The United Nations Food and Agriculture Organization is providing leadership for sustainable agriculture.

12.1 Introduction

There are many topics related to reducing greenhouse gas emissions that are associated with agriculture in particular. Agriculture connects closely with sustainability issues having to do with land use, water use, and food security/world hunger issues. At the same time, there are other issues, already discussed more generally, that are relevant for agriculture as well: Renewable energy generation, electrification of transportation (here including farm machinery), and a smarter infrastructure.

The generation of electricity using wind and solar energy for farming with electric tractors has a bright future. The cost of solar panels continues to decrease, and battery storage of renewable electricity has become competitive with other alternatives (see Chapter 8). The progress in battery developments is very important for agriculture because battery purchase prices below $50/kWh, combined with decreasing costs for electricity from wind and solar energy, will bring low energy prices to agriculture.

The electrification of transportation within agriculture includes electric tractors and trucks, as well as cars. There is progress with respect to the development of new products and their use by farmers. The electrical energy for use on the farm can be produced by adding solar panels to farm buildings and generating electricity for local use, and battery storage can be used to manage supply and demand of electricity. Some farms are able to be connected to a local electrical grid, whereas others may need to be served by an off-grid system developed and operated by the farmer (see Chapter 9).

There are land use issues that arise in agriculture because of the competition between food production, crops intended for ethanol production, land devoted to wind and solar energy generation, and land used for the production of wood (paper) and cotton (fabrics). Some of these land use issues are less difficult than others. For instance, solar panels can often be placed on buildings, over parking lots, and on land that is not suitable for farming.

Much of the technical development for advancing sustainable agriculture has been achieved, and part of the challenge now is getting those developments recognized as having great value. Some of the acceptance process in agriculture relies on what farmers have learned by experimenting in their fields. Therefore, it is important that extension service professionals continue to distribute this valuable information. The Internet has also, of course, become a valuable library and information system that is being used in support of agriculture.

12.2 Sustainable Development Goals

As discussed in Chapter 2, the UN General Assembly (United Nations, 2015) adopted the resolution *Transforming Our World: The 2030 Agenda for Sustainable Development* on September 25, 2015. This resolution includes a set of 17 Sustainable Development Goals (SDGs) that are very high priorities for both the United Nations and the signatories of the Paris Climate Agreement. Many of these SDGs relate clearly to agriculture.

The second of these SDGs is to end hunger, achieve food security and improved nutrition, and promote sustainable agriculture. The sixth SDG includes sustainable management of water. The seventh SDG is to ensure access to sustainable energy for all. The fourteenth SDG includes sustainable use of marine resources. The fifteenth SDG promotes sustainable management and use of land and terrestrial ecosystems. The Food and Agriculture Organization (FAO) has identified principles and 20 actions in support of the 2030 Agenda for Sustainable Development (FAO, 2018). These 20 actions are strongly interrelated and are described as follows:

Action 1: Facilitate access to productive resources, finance and services. Rural infrastructure, resources and services are important to support agriculture, and renewable electricity needs to be a part of rural infrastructure. The decrease in costs of solar panels and batteries is beneficial to agriculture, including those who must develop their own off-grid generating system.

Action 2: Connect smallholders to markets. There is significant value in having an infrastructure that allows those in agriculture to connect with people in communities that provide agricultural supplies and consumers of agricultural products (markets). Inexpensive electric vehicles (EVs) powered by solar energy can help rural residents connect to local communities in an inexpensive and sustainable way.

Action 3: Encourage diversification of production and income. Cultivating multiple crops has value for many producers and increases resilience in the face of weather, disease, or other adverse events. Additionally, having diverse income streams can mean employment beyond the farm for one or more members of a family, providing opportunities that often benefit the family. Again, an inexpensive EV has value for those seeking diversification.

Action 4: Build producers' knowledge and develop their capabilities. Education has great value for farmers, and there are many opportunities to learn from extension organizations, farm organizations, and others. The Internet has many sites that are beneficial for producers. It is beneficial for producers to have electricity, to be connected to the Internet, and to be able to have access to market information and weather forecasts.

Action 5: Enhance soil health and restore land. Any good farmer will tell you that good soil stewardship is necessary for success. Traditionally this has largely meant optimal use of fertilizer, but that is now just one aspect of soil management. Because of climate change, soil erosion associated with intense rainfall and flooding has become a greater problem than in the past. The FAO has voluntary guidelines for sustainable soil management (FAO, 2017) that include actions to minimize soil erosion, enhance soil organic matter content, foster soil nutrient balance, minimize soil salinization, acidification, and alkalinization, prevent soil contamination, enhance soil biodiversity, reduce soil compaction, and improve soil water management (FAO, 2017).

Action 6: Protect water and manage scarcity. Water quantity and water quality are central topics in agriculture because about 70% of all water is used for crops and livestock. Insufficient or polluted water can thus have tremendous impacts on food products, which then also have impacts on consumers' health.

Water scarcity is an issue in many locations where crop irrigation is an active process, using wells, canal systems, or both. As water becomes less prevalent, efforts to improve water efficiency will have more value and should be extended with further developments. Sustainable management of water resources more broadly is also important for society; it is needed to end hunger and it is the sixth SDG.

Action 7: Mainstream biodiversity conservation and protection of ecosystem functions. Healthy ecosystems are a valuable part of efforts to achieve food security and sustainable agriculture. This can include protecting not only wildlife in general, but also ecosystem biodiversity in places such as soil microorganisms. Healthy and diverse organism populations in soil are generally beneficial, and biodiversity in ecosystems has value in efforts to increase food production. International progress in programs to maintain important genetic resources for plant agriculture has been reported (FAO, 2015), and many programs exist to maintain genetic resources and to enable good ecosystem function in agriculture in many locations.

Action 8: Reduce losses, encourage reuse and recycling, and promote sustainable consumption. The issue of managing food products so that they are used effectively and efficiently is a substantial challenge because many foods may spoil. Somewhere around one-third of food production is lost due to poor food management. Packaging, storage conditions, managing supply and demand, preservation methods, and consumer education are aspects of programs to reduce food waste (FAO, 2011; Esguerra, 2018). Good management of food products has significant economic value and needs to be an important priority.

Action 9: Empower people and fight inequalities. One way to empower people is to educate them. Education is very important in agriculture because knowledge regarding farming practices, product management, renewable energy choices, water management, and soil stewardship are foundational to implementing these practices, and therefore these educational efforts have great value. Children that live in rural areas, of course, should go to school and receive an education. Part of that education should include information tailored to the situations those children are likely to encounter as they grow up and begin their adult careers.

Action 10: Promote secure tenure rights. The right to have access to land to farm it is an issue in some locations. Land ownership is a key factor because soil stewardship is often better when the person managing the land is the owner; they have more of an incentive to improve soil quality. One of the issues in ocean stewardship is the problem of managing the ocean as a common resource (see Chapter 15).

Action 11: Use social protection tools to enhance productivity and income. Agriculture is a physically, cognitively, and socially demanding occupation. There are many risks for those who are working in production agriculture. Social protection programs reduce risks and have long-term value in agriculture. For instance, food availability for all has been demonstrated to work effectively and have positive results.

Action 12: Improve nutrition and promote balanced diets. Nutrition education is beneficial because it allows people to understand the health values associated with food choices. Having the opportunity to select nutritious foods for a diet is one part of the food pyramid, along with education about what food choices are better than others. The affordability of milk, meat, fruits, and vegetables impacts health for some because they do not have sufficient resources to purchase food for a good diet.

Action 13: Prevent and protect against shocks to enhance resilience. There are a number of national disaster risks associated with agriculture: Floods, wildfires, and very dry conditions (including droughts). Resilience is important in agriculture because disasters such as these recurrently happen and those impacted often need help. Furthermore, because of climate change, there have been more of these types of disaster events recently. Good communication systems are valuable when timely information is needed.

Action 14: Prepare for and respond to shocks. There is significant value in having prepared for potential disasters, having emergency supplies, and having a plan to follow to address an emergency. Even basic and effective communication with respect to impending weather and wildfire events has great value and can save lives or livelihoods.

Action 15: Address and adapt to climate change. As mentioned before, agriculture is being impacted by climate change – in some ways more severely than other sectors of the population – in the form of more floods, wildfires, and droughts than in earlier years. The severity of these events is also exacerbated. This feeds back into the need to reduce greenhouse gas emissions associated with agriculture (for example, by using electricity from renewable sources, electrifying operations, and producing fertilizer with sustainable energy and electrolysis of water).

Action 16: Strengthen ecosystem resilience. Whether ecosystems are functioning, faltering, or failing, they are impacted by weather and climate change. Actions are often needed to restore ecosystems and improve their health and resilience after disasters strike. This includes updated evaluation methods that allow better assessments of lands and ecosystems to determine what actions are needed to improve the state of the land at the site of interest.

Action 17: Enhance policy dialogue and coordination. Government policy has been important in agriculture in many countries for many years (see Chapter 10). Research, education, and extension programs have provided support for agriculture since 1863 in the United States. There have been global cooperative efforts through the FAO for many years that have been very beneficial. Policy actions are almost always an important part of efforts to advance sustainable agriculture and reduce greenhouse gas emissions. Of particular importance currently is a more extensive dialogue on water policies in support of agriculture using renewable energy.

Action 18: Strengthen innovation systems. Innovation has always been very important in the success of agriculture in the past, and it is needed moving forward into the future to achieve the SDGs. The transformation to renewable energy, the electrification of agricultural operations, and the development of battery storage systems to support rural electrification all depend on further innovation. Innovation is also important in sustainable ecosystem management.

Action 19: Adapt and improve investment and finance. Investments in agriculture are necessary for reaching the SDG to end hunger and achieve food security for everyone. Land ownership is a large investment, as are the investments in farm equipment, seeds, and fertilizers that are needed for modern agriculture. These all require working capital. Additional investment is needed for solar panels and battery storage to have renewable electricity. A modern home with appliances and a communication network completes the picture of what is desired. A nearby community with schools, a hospital, stores, a bank, a post office, and a local government should be part of the rural infrastructure.

Action 20: Strengthen the enabling environment and reform the institutional framework. Rural development in support of sustainable agriculture is important. Support and policies implemented by local and state governments are significant, but other organizations are involved as well in many countries. For instance, there are global companies that have products such as tractors and farm equipment for sale, and these companies can play key roles in the electrification of that equipment. Education and the availability of information, naturally, should also be part of the enabling environment.

12.3 Greenhouse Gases and Agriculture

Agriculture has several important impacts on greenhouse gas emissions, some of which are more obvious than others. Equipment, transportation of products, and nitrogen-based fertilizers contribute to emissions. But organic carbon in soils is also beneficial in many locations because crop yields are better in soils with higher concentrations of organic matter and good biodiversity. Thus, actions by farmers to increase soil organic matter and water infiltration into soils can often have positive results in environments where increasing soil moisture availability has benefits. Another less obvious implication for greenhouse gas emissions from agriculture is the potential for ocean-based farming.

The oceans have the potential to support the growth of algae, phytoplankton, and other photosynthetic microorganisms that fix carbon. One of the opportunities that exists is to enhance biomass production in the oceans and other locations. Food production in water environments could be increased substantially. There certainly are science and technology development challenges to farming the oceans, but there are also great opportunities to increase production of food from the oceans. Additionally, there are important social and policy issues to develop and put in place in order to allow ocean mariculture farmers to lease a plot in the ocean and responsibly develop it into a more productive ecosystem (EU, 2017). Progress is needed in managing the commons of the ocean and investing in developing seafood farming technologies for ocean environments. Increasing photosynthetic production in the ocean, however, can increase the harvest of useful products that are higher up on the food chain.

One option, for illustrative purposes, would be to develop a community of people living in houseboats on the ocean at a selected location with an approved global position and an approved geographical area to use productively. Minerals and fertilizers that are needed for primary production of algae and phytoplankton could be added to the water regularly to support photosynthetic growth. The organisms needed to have a productive ecosystem could be added at the site to develop a productive community. Seaweed and other plant communities could be established. Solar panels with battery storage could be used for electricity. The community could have a boat to take harvested products to market and bring supplies from land regularly. Delivery using drones might be possible as well. There is a large part of the ocean that needs to be fertilized to have productive primary growth of biomass. Nitrogen, phosphorus, iron, and silicon are the elements that have been identified as limiting nutrients for primary productivity in the ocean (Sigman and Hain, 2012). While about 80% of the surface of the Earth is covered by water, only about 50% of primary carbon fixation by photosynthesis occurs in the oceans.

This is not a far-off hypothetical scenario. There are some individuals that are farming the ocean and producing seaweed, kelp, and other sea vegetables using a three-dimensional underwater scaffolding support system (Schiffman, 2016). Oysters, mussels, scallops, and shellfish are harvested in the area of the growing plants. One of the challenges of ocean farming, though, is to manage the ocean farming volume in an environmentally positive way such that environmental quality improves – rather than degrades – because of the actions taken to make the location productive.

12.4 Renewable Fuels

Some renewable fuels are produced from crops and crop waste materials. Perhaps most notably, ethanol is produced in significant quantities in several countries. The most common substrates for ethanol fermentation in the United States are corn and grain sorghum. Sugarcane is a common substrate in Brazil, which produces ethanol in significant quantities. Cellulose has been investigated as a substrate, but the economics are not very good at the present time (2019). Waste carbohydrate sources, such as grain dust cereal wastes, can also potentially be used to produce ethanol.

Another renewable fuel option from agriculture is through anaerobic digestion of agricultural wastes to produce methane. Waste products with this potential include dairy manure, food manufacturing waste, feed waste, and food waste. The economics of methane production depend on the alternatives that are available, but where methane from natural gas is expensive, the process of anaerobic digestion can provide gas that has value as a renewable fuel.

12.5 Air Quality

Finally, the issue of good air quality can centrally include agricultural concerns. The burning of crop residues and pastures impacts air quality significantly where it is allowed. The use of solid fuels and fires for cooking and heating, common in underdeveloped rural areas that often overlap with agricultural areas, causes air pollution and impacts health.

Larger domesticated animal operations have a different set of air quality issues that can arise. Dust from large cattle (and other animal) feeding operations is an issue when the feedlot soil is dry. There are odors from animal feeding operations in many cases. Where animal feeding takes place inside buildings, indoor air quality may be a problem. There can also be an issue with microbial pathogens that may move through the air to contaminate other organisms.

References

Esguerra, E.B. (2018). *Case studies on managing quality, assuring safety and reducing post-harvest losses in fruit and vegetable supply chains in South Asian Countries*. UN Food and Agriculture Organization; www.fao.org/publications

EU (2017, November 29). *Food from the Oceans*. European Commission, Brussels, Belgium.

FAO (2011). *Global Food Losses and Food Waste*. UN Food and Agriculture Organization; www.fao.org/publications

FAO (2015). *Guidelines for Developing a National Strategy for Plant Genetic Resources for Food and Agriculture*. UN Food and Agriculture Organization; www.fao.org/publications

FAO (2018). *Transforming Food and Agriculture to Achieve the SDGs*. UN Food and Agriculture Organization; www.fao.org/publications

FAO (2017). *Voluntary Guidelines for Sustainable Soil Management*. UN Food and Agriculture Organization; www.fao.org/publications

Schiffman, R. (2016). Can we save the oceans by farming them? *Yale Environment* 360, Yale University, New Haven, Connecticut; https://e360.yale.edu/

Sigman, D.M. and Hain, M.P. (2012). The biological productivity of the ocean. *Nature Education Knowledge* 3(10): 21.

United Nations (2015). *Transforming Our World: The 2030 Agenda for Sustainable Development*. United Nations, New York.

13

International Developments

Abstract

The challenge of reducing greenhouse gas emissions is a global one, and it is important to make efforts to decrease combustion processes in all countries. At the same time, it is also important to take into consideration the particular circumstances of different areas, because the specific needs and opportunities can vary by region. Improving air quality has local and regional challenges and it is possible for a metropolitan area to make significant progress towards reducing emissions in a community. This chapter focuses on three locations – China, London (UK), and Norway – as examples of how to approach this global challenge in different, local ways. China is a major source of greenhouse gas emissions, and there has been a recent effort to improve air quality in many large cities. Central London is a single city, designated as a low emission zone in 2008, and since then there has been an effort to improve air quality in central London. Norway is a lower-population country that has had an ongoing effort to reduce greenhouse gas emissions and improve air quality for several years, and has several incentives to encourage citizens to purchase electric vehicles. Improving air quality is an important goal in all three cases, even though their approaches have varied. Norway has made much greater progress than China and London in reducing greenhouse gas emissions.

13.1 Introduction

There are significant developments underway in many countries related to reducing greenhouse gases and improving air quality. Because of the more immediate and direct impact of air quality on health, there is considerable information on regulations and management efforts to improve urban air quality (Gulia et al., 2015). Many countries have urban air quality management plans, with regulations and monitoring of important pollutants such as particulate matter, nitrogen oxides, and ozone.

Locations around the globe differ, however, along a number of dimensions that are relevant to how best to approach improving air quality and reducing greenhouse emissions. In particular, countries vary in their overall land size, population and population densities, degree of industrialization, government structure and authority, and some of the inherent options available in the local environment (e.g. hydroelectric power).

One of the most important countries to include in this chapter is China. China is both geographically large and also has a very large population, causing it to have a correspondingly high amount of emissions and negative health impacts. It is also a developing country in several respects, and it has a relatively strong central government that can enact large-scale policies.

The United Kingdom is geographically much smaller, but there is an interesting case to focus on in the city of London. London is a major city, in a developed country, with significant air pollution problems because of nitrogen oxide emissions from motor vehicles. As a single city with a more specific issue, London is able to take more particular actions to address that issue.

Norway is a developed country with a smaller population than the city of London (but it is geographically about the same size as the entire UK), and it has been leading the way in taking steps to reduce greenhouse gas emissions and improve air quality. In particular, Norway is the leading country with respect to the electrification of transportation.

13.2 China

In China, approximately 1.1 million people died prematurely in 2015 because of outdoor ambient air pollution (Huang et al., 2018), and about 0.8 million died prematurely in 2013 because of direct household exposure to solid fuels (Liu et al., 2016). There are significant efforts in progress to reduce greenhouse gas emissions and improve air quality. Because of the very significant and immediate health impacts of air pollution, improving air quality has a higher priority than reducing greenhouse gas emissions in China. Typically, both air quality and carbon emissions are strongly interlinked, but there are exceptions to this relationship, and for this region the air quality issues are predominant.

In 2015, 64% of primary energy supply in China was from coal (Qin et al, 2018). Approximately 75% of the electricity that was generated in 2015 was from coal-fired power plants (Peng et al, 2018). Approximately 14–17% of electricity was from hydropower (Lam et al., 2016). Since natural gas is a cleaner fuel, there is an effort in progress to increase natural gas use from 6% to 10% of primary energy supply by 2020. One way this is being accomplished is by converting coal to synthetic natural gas (SNG).

There has been significant progress in improving air quality under the 2013 Air Pollution Prevention and Control Action Plan (APPCAP). Between 2013 and 2017, PM2.5 concentrations decreased by about 33%, and sulfur dioxide concentrations decreased by about 54%; these changes were significant. On the other hand, changes in concentrations of nitrogen oxides were small, and ozone concentrations increased (Huang et al., 2018). Because of the importance of PM2.5 with respect to health, there were 47,240 fewer deaths in 2017 compared to 2013 in the 74 key cities that were included in the study (Huang et al., 2018). During this time period, there have been ongoing efforts to reduce emissions by electrifying transportation, generating more electricity with wind and solar energy, and using other aspects to increase clean energy supply. There has also been some transition from coal to natural gas for the generation of electricity, and electricity generation with wind and solar energy has advanced significantly.

These results may be compared to the predicted results based on the APPCAP. For PM2.5, the simulated results predict a decrease of 30% by 2017 and 40% by 2020 for the Beijing-Tianjin-Hebei Region (Cai et al., 2017). These values are in reasonable agreement with the measured values of Huang et al. (2018) for 2017 for the 74 key cities. However, the simulated results show decreases of 31% in 2017 and 44% in 2020 for nitrogen oxides, while the measured values show a 9.7% reduction for 2017.

In the effort to improve air quality, there has been a residential transition from solid fuels to gas for heat and cooking, which resulted in a decrease of 0.36 million premature deaths per year from 2005 to 2015. It is estimated that the integrated population-weighted exposure (IPWE) to PM2.5 has decreased by 47% from 2005 to 2015 (Zhao et al., 2018). This IPWE includes both ambient air pollution and household air pollution exposures.

Why have there been areas of uneven progress within China, particularly with reference to nitrogen oxides and ozone? Part of this is because one of the pathways to increase gas for cooking and heating is to produce synthetic natural gas (SNG) from coal. When SNG is used by residential households for heating and cooking, this provides the greatest air quality and health benefits when it replaces solid fuels such as coal (Qin et al., 2017; Qin et al., 2018). This improves air quality, but it also increases greenhouse gas emissions.

In 2018, the Chinese government announced a new three-year effort to further reduce air pollution (Greenbaum, 2018). The efforts to electrify transportation are continuing in China with increasing sales of electric vehicles including cars, buses, taxis, and electric bicycles. The electrification of transportation is one part of the effort to improve urban air quality.

China has been one of the major developers of solar photovoltaic technology to generate electricity using solar radiation. There is significant progress being made in China to install solar panels and generate electricity. More than 100 GW had been installed as of 2018 (Yang et al., 2018). The goal for 2030 is 400 GW, which will be about 14–16% of the 2500–2800 GW of installed electric generating capacity in 2030. This solar-generated electricity has two benefits: Air quality is better and there are no greenhouse gas emissions associated with solar photovoltaic operations. Solar panels manufactured in China are also competitive in the global market.

In 2016, China had 168.7 GW of installed wind power capacity and generated 241 TWh of electricity (Huenteler et al., 2018). Two challenges that have reduced the amount of electricity that is used are connecting the new wind power to the grid and curtailing the electricity flow to users because of grid capacity and grid management (Lam et al., 2016). There is the potential to install 2400 GW of wind capacity by 2050 and produce 5350 TWh of electricity per year, which would be about 35% of the electricity generated (Huenteler et al., 2018). One of the important issues in China is to improve site selection.

There are also many other electric power issues in China that are being considered as the efforts to provide power, improve air quality, and reduce greenhouse gas emissions continue. Market reforms that are designed to reduce the cost of electricity to consumers include the introduction of wholesale and retail competition (Lin et al., 2019). Planning studies on carbon dioxide emission pricing and its impact on electricity are considered in this study. Since coal is one of the lower-cost options in China, a high carbon dioxide price is needed to alter the choice of sources of power. Both PM2.5 concentrations and carbon emissions are affected by generating electricity using coal.

Simulation has been used to investigate alternatives and the impacts on air quality, health, and carbon emissions in 2030 (Peng et al., 2018). If the 2015 state of 75% electricity from coal is maintained, there are air quality and health benefits from transitioning to 30% electric vehicles and 30% of the residential population shifting from solid fuels to electricity for residential heating and cooking. There are even greater benefits if only 50% of the electricity comes from coal. The avoided deaths in China are 55,000 to 69,000 when only 50% of the electricity is generated using coal, compared to 41,000 to 57,000 for 75% of electricity from coal. Shifting to only 50% coal for power generation results in a 14–16% reduction in carbon emissions.

13.3 London, United Kingdom

London is the largest city in the United Kingdom, the seat of government, and the cultural center of Britain. In contrast to China, it is in a fully developed country and it is geographically very small. A major issue in London is the specific need to reduce nitrogen oxide emissions. Poor air quality is the greatest environmental risk to public health in the United Kingdom (Joint Air Quality Unit, 2017), and there is an ongoing effort to take action to improve air quality in London. Between 2010 and 2015, nitrogen oxide emissions were reduced by 19%. In 2013, about 57% of nitrogen oxide emissions were from road transportation, and 45% of these were associated with buses and coaches, 18% with trucks, and 15% with taxis (Simons, 2019). One of the actions that has contributed to a reduction in nitrogen oxide emissions is progress in motor vehicle pollution control associated with regulations that require reduced emissions (Font et al., 2019).

Central London has had a low emission zone since February 2008 (Font et al., 2019). This zone has had greater success in reducing particulates associated with transportation, rather than nitrogen oxides (Font et al., 2019). On April 8, 2019, a new Ultra Low Emission Zone was established for central London, which is an area slightly larger than one square mile. Cars, vans, and motorcycles that emit nitrogen oxides and do not meet the emission standards have to pay £12.50 daily to drive into central London, in addition to the £11.50 congestion charge (Badstuber, 2019). For buses and lorries (trucks) that do not meet emissions standards, the charge is £100/day.

The goal is to have over 90% of central London meet the World Health Organization (WHO) guidelines for nitrogen oxides by the beginning of 2025. Values are presently significantly above the value of 40 micrograms per cubic meter, which is the WHO guideline for the annual mean value. The Greater London urban area has the highest concentrations of nitrogen oxides, but there are also other cities in the UK that are not meeting the WHO guidelines. Another goal is to have almost all cars and vans be zero-emission vehicles by 2050 (Joint Air Quality Unit, 2017).

There are now regulations on taxis. New taxis must be zero-emission capable. New private hire vehicles will have zero-emission capable requirements phased in starting in 2020 (Simons, 2019). For London taxis, there is a 15-year age limit. New electric taxis are being manufactured in the UK and being used in London.

For buses, there is an ongoing effort to introduce new zero-emission buses, and a goal of having all single-decker buses be zero emission by 2020. There is also a transition to zero-emission double-decker buses (Simons, 2019). Since buses are a major source of nitrogen oxide emissions, moving forward with zero-emission buses will be very beneficial to efforts to improve air quality.

There is a proposal to transition to a central London Zero Emission Zone by 2030, with at least 90% of all vehicles being zero-emission vehicles (Simons, 2019). Some streets will be for walking, with no vehicles permitted at some times. The effort to increase the number of zero-emission vehicles has started and is continuing. There is a program to encourage the addition of electric vehicle (EV) charging infrastructure, as well as support for research and development at universities in London (Simons, 2019). The City of London is also moving forward with zero-emission vehicles for waste collection and street cleaning. There is a goal to have all city-owned and -leased vehicles be zero emission or hybrid by 2025.

London has a program to increase transportation by bicycle, and this resulted in a 292% increase in cycling from 1999 to 2017 (Simons, 2019). Parking for bicycles and services in support of cycling have correspondingly been increased.

Since the City of London central area is slightly more than one square mile, it is very easy to walk to destinations in this area. Thus, arriving by bus or train and then walking to a final destination works well for many people. There is an ongoing effort to improve the conditions for walking, including improving air quality.

Trucks that need to deliver products to the City of London can do so without emissions if the trucks are zero-emission vehicles. One of the ongoing concerns with diesel trucks is idling. This problem can be eliminated by restricting diesel trucks from entering the City of London.

Finally, emissions from buildings have been identified because they impact air quality and the effort to reduce carbon emissions. Buildings owned by the city are included in the air quality improvement plan. There is an educational effort to enable facilities managers to reduce emissions.

The London Local Air Quality Management plan includes continuing efforts to monitor and assess air pollution by having Air Quality Management Areas wherever regulations are not being met, implementing and following the current action plan, updating the plan every five years or less, and publishing annual reports on progress (Simons, 2019). Although central London is just one city, it is shaping up as a key proving ground for many programs to reduce air pollution and greenhouse gas emissions. Effective methods that have been worked out in London can then provide a set of blueprints for other cities in the United Kingdom and beyond.

Children's respiratory health is one of the significant concerns in urban areas with poor air quality. The impact of the low emission zone in central London on children's lung health was investigated from 2009 to 2014. Lung volume is smaller in children that are impacted by high nitrogen oxide concentrations and high concentrations of particulates (PM2.5 and PM10). Forced vital lung capacity was found to be inversely proportional to concentrations of nitrogen oxides and particulates (Mudway et al., 2019). However, there was not a significant correlation between forced expiratory lung volume and nitrogen oxide concentrations. There was no evidence of a decrease in the proportion of children with small lungs over the time period of the experiment (Mudway et al., 2019).

13.4 Norway

Norway is a small country with about 5 million people that is fully developed and has some inherent options for renewable energy due to its location. It is also a country that has done far more than just take advantage of those natural options. Norway has implemented a number of incentives to reduce greenhouse gas emissions and improve air quality.

Oslo is the largest city, with about 670,000 people, and it has the highest concentrations of PM2.5 and nitrogen oxides (NEA, 2018). The most important emissions are from transportation in Norway. The pollutants PM2.5 and nitrogen oxides are of greatest concern (UNEP, 2016). The incentives to purchase zero-emission vehicles include the use of bus lanes, toll-free travel, privileged parking (Hockenos, 2017), and exemptions from sales, import, and road taxes (Chappell, 2019). Wood burning for heating is an important source of emissions in the winter (NILU, 2017).

Norway is currently the leading country in terms of the percentage of automobiles that are plug-in electric vehicles, and, per capita, Oslo has the most electric vehicles in the world (UN, 2018). In March 2019, 58% of new car sales were EVs (Chappell, 2019). Because of the success of incentives to drive EVs, greenhouse gas emissions have decreased by 35% since 2012, as of 2018, with a goal of 95% reduction by 2030 (UN, 2018). In Norway, there is great progress being made in reducing greenhouse gas emissions and improving air quality (UN, 2018; NEA, 2018). Both PM2.5 and nitrogen oxides in urban cities in Norway have declined because of the increased percentages of EVs in Norway (NEA, 2018).

In Norway, most of the electricity is generated with hydropower. There are about 1000 storage reservoirs in Norway, and the amount of power generation can be increased or decreased to meet demand. There is also power generation associated with river flow, which cannot be increased or decreased to the same extent as that in reservoirs (EF, 2019). As of early 2018, Norway had 1660 hydropower plants, with 31,837 MW of installed capacity, which is 96% of all installed capacity. About 75% of production capacity is flexible, and Norway has been an exporter of electricity for much of the time. In 2016, about 10% of the generation (16.5 TWH) was exported (IHA, 2017).

At the end of 2017, Norway had 33 wind farms with 1188 MW of installed capacity, and additional wind farms were under construction (EF, 2019). There were 32 thermal power plants as of 2017, with 1108 MW of installed capacity. Fuels include municipal and industrial waste, natural gas, oil, and coal (EF, 2019).

Norway has electric buses, electric taxis, and plans for an all-electric ferry (Manthey, 2018; Hanley, 2019). There are goals to have all public transportation be powered by electricity and for Oslo to be a zero-emissions city by 2030 (Hanley, 2019). All taxis are to be zero-emission vehicles in the future as well.

References

Badstuber, N. (2019, April 8). London's new charge on polluting vehicles. Here is everything you need to know. World Economic Forum; www.weforum.org/

Cai, S., Wang, Y., Zhao, B. et al. (2017). The impact of the "Air Pollution Prevention and Control Action Plan" on PM2.5 concentrations in Jing-Jin-Ji region during 2012–2020. *Science of the Total Environment* 580: 197–209.

Chappell, B. (2019, April 2). *Electric cars hit record in Norway, making up nearly 60 percent of sales in March*. National Public Radio; www.npr.org/
EF (2019). *Energy Facts*. Norway; energifaktanorge.no/
Font, A., Guiseppin, L., Blangiardo, M. et al. (2019). A tale of two cities: Is air pollution improving in Paris and London? *Environmental Pollution* 249: 1–12.
Greenbaum, D. (2018). Making measurable progress in improving China's air and health. *Lancet Planet Health* 2: e289–e290.
Gulia, S., Nagendra, S.M.S., Khare, M., and Khanna, I. (2015). Urban air quality management – A review. *Atmospheric Pollution Research* 6: 286–304.
Hanley, S. (2019). Oslo adding 70 electric buses this year. *Clean Technica*; cleantechnica.com/
Hockenos, P. (2017). With Norway in the lead, Europe set for surge in electric vehicles. *Yale Environment* 360; e360.yale.edu/
Huang, J., Pan, X., Guo, X., Li, G. (2018). Health impact of China's Air Pollution Prevention and Control Action Plan: An analysis of national air quality monitoring data. *Lancet Planet Health* 2: e313–e323.
Huenteler, J., Tang, T., Chan, G., and Anadon, L.D. (2018). Why is China's wind power not living up to its potential? *Environmental Research Letters* 13: 044001.
IHA (2017). *Norway Statistics*. International Hydropower Association; www.hydropower.org/
Joint Air Quality Unit (2017, May). Improving air quality in the UK: Tackling nitrogen dioxide in our towns and cities. Department for Transport, UK; www.gov.uk/defra
Lam, L.T., Branstetter, L., and Azevedo, I.M.L. (2016). China's wind electricity and cost of carbon mitigation are more expensive than anticipated. *Environmental Research Letters* 11: 084015.
Lin, J., Kahrl, F., Yuan, J. et al. (2019). Economic and carbon emission impacts of electricity market transition in China: A case study of Guangdong Province. *Applied Energy* 238: 1093–1107.
Liu, J., Mauzerall, D., Chen, Q. et al. (2016). Air pollutant emissions from Chinese households: A major and underappreciated ambient pollution source. *Proceedings of the National Academy of Sciences* 113: 7756–7761.
Manthey, N. (2018). VARD to build all-electric ferry for Boreal in Norway; www.electrive.com/
Mudway, I.S., Dundas, I., Wood., H.E. et al. (2019). Impact of London's low emission zone on air quality and children's respiratory health: A sequential annual cross-sectional study. *Lancet Public Health* 4: e28–e40.
NEA (2018). *Air Pollution*. Norwegian Environment Agency; www.environment.no/
NILU (2017). *Wood burning pollutes the urban air in Norway*. Science Nordic, Norwegian Institute for Air Research; sciencenordic.com/
Peng, W., Yang, J., Lu, X., and Mauzerall, D.L. (2018). Potential co-benefits of electrification for air quality, health, and CO2 mitigation in 2030 China. *Applied Energy* 218: 511–519.
Qin, Y., Hoglund-Isaksson, L., Byers, E. et al. (2018). Air quality-carbon-water synergies and trade-offs in China's natural gas industry. *Nature Sustainability* 1: 505–511.
Qin, Y., Wagner, F., Scovronick, N. et al. (2017). Air quality, health, and climate implications of China's synthetic natural gas development. *Proceedings of the National Academy of Sciences* 114: 4887–4892.

Simons, J. (2019, March). *City of London Air Quality Strategy*. City of London; www.cityoflondon.gov.uk/air

UN (2018). Oslo takes bold steps to reduce air pollution, improve livability. UN Environment, United Nations Environmental Program; www.unenvironment.org/

UNEP (2016). *Norway air quality policies*. UN Environmental Program; wedocs.unep.org/

Yang, J., Li, X., Peng, W. et al. (2018). Climate, air quality and human health benefits of various solar photovoltaic deployment scenarios in China in 2030. *Environmental Research Letters* 13: 064002.

Zhao, B., Zheng, H., Wang, S. et al. (2018). Change in household fuels dominates the decrease in PM2.5 exposure and premature mortality in China in 2005–2015. *Proceedings of the National Academy of Sciences* 115: 12401–12406.

14

Examples of Progress

Abstract

Many places in the world have made significant progress towards reducing greenhouse gas emissions and improving air quality. These places show that reducing combustion has significant social, environmental, and economic benefits for those locations. Recent analyses that include both climate change impacts and the effects of air quality on health show that it is beneficial to replace coal-fired electricity generation with wind and solar power. This chapter highlights some of the advances in several of these areas. Cost/benefit analysis shows that electric buses, cars, and trucks have features that are beneficial to society and appreciated by riders and those living nearby. Both the costs of electricity generation with wind and solar systems and the costs of battery storage of energy have decreased in the last ten years. As of 2019, it is desirable to move steadfastly forward with actions that reduce greenhouse gas emissions and improve air quality.

14.1 Introduction

There are many examples of progress that could be included in this chapter to illustrate that the transitions towards reduced greenhouse gas emissions and improved air quality are moving along in many parts of the world. In this chapter, some recent information is presented on electric vehicles, battery storage of electricity, and the co-benefits of considering both greenhouse gas emissions and air quality.

14.2 Electric Vehicles

There continues to be considerable excitement about the development of electric vehicles (EVs). Although some of it may be exaggerated, it is definitely the case that most of that excitement is justified. Personal transportation is a significant source of both poor air quality and greenhouse gas emissions (particularly if one considers the entire petroleum extraction, refinement, distribution, and combustion process). The emission from vehicles, furthermore, are particularly concentrated in larger cities, where they influence a disproportionate number of other people in the population. And, of course, there are economic and personal value reasons why individuals can choose to participate in the transition to EVs. Thus, there is huge potential for continued progress with regard to EVs.

14.2.1 Electric Buses

One of the areas for developments with the most significant potential for progress is the electric bus. As of May 2019, China had about 421,000 electric buses, Europe about 2250, and the United States about 300 (Eckhouse, 2019). Barcelona, Spain has a project to replace diesel buses to reduce pollution in the city; the plan includes 116 new electric buses and some hybrid buses to be delivered in 2019, 2020, and 2021 (Dzikiy, 2019). The buses are considered to be safer and less polluting. Paris, France has a plan to replace diesel buses with 800 new electric buses (Dzikiy, 2019).

The transition to electric buses is not just happening in national capitals or other major cities, though. Electric buses are showing up in smaller communities such as Asheville, North Carolina, which put five new electric buses into service in June 2019 as part of an effort to reduce the city's carbon emissions (Burgess, 2019).

Electric buses are used for many hours each day, and the estimated savings in diesel fuel are 270,000 barrels per day in 2019 for the buses that are in service (Prosser, 2019). The transition to electric buses is very beneficial to society because of the improved air quality and the reduction in greenhouse gas emissions. Better air quality is one of the reasons that there has been so much progress in the transition to electric buses.

14.2.2 Electric Cars

Electric cars are becoming increasingly established and the electric car industry is now rapidly expanding. Electric cars are continuing to evolve, with many new models being introduced. Volkswagen has plans to have 70 models and build 22 million EVs by 2030 (Evarts, 2019d). Batteries are important for EVs, and Volkswagen has four companies to supply batteries. Audi has reported on its plans to introduce 20 new battery electric models and ten plug-in hybrid electric models by 2025 and to have its percentage of EV sales increase to 40 (Evarts, 2019a). BMW plans to have more than ten plug-in hybrid and electric models on sale by the end of 2020, with 12 all-electric models and 13 plug-in hybrids by 2025 (Evarts, 2019b). BMW plans to spend about $56 million to expand its battery factory to produce advanced fifth-generation battery packs. The efforts include research and development to produce new battery cells from recycled materials from old battery packs.

More than 40 EVs were on display at the 2019 New York auto show, including the Mercedes-Benz EQC, Volkswagen ID Buggy, and 2020 Lincoln Corsair (Eisenstein, 2019). Also at the auto show was a new next-generation electric vehicle charger that can deliver 350 kW, which provides about 20 miles of range per minute of charging. The significant message for those who viewed the new models is that customers have many more choices in 2019 and there will be even more in 2020.

Sales of EVs in Europe were up in the first quarter of 2019, with about 125,400 deliveries, and total EV sales exceeded 100,000 in a quarter for the first time (Gaton, 2019). One of the factors that increased demand was the availability of the Tesla Model 3. In Norway, 59% of new car deliveries were EVs compared to 13.5% in Sweden. Plug-in electric sales were up 31%, even though battery delivery bottlenecks reduced the availability of some models.

14.2.3 Electric Trucks

In contrast to electric cars, there is very little currently available in the form of electric trucks. The technology from electric cars can transfer to trucks with relatively minor adjustments, though, and this seems to be a significant area for growth. There is great interest in a new company, Rivian in Plymouth, Michigan, which has developed an electric pickup truck, R1T, that has a front trunk, fast acceleration, the ability to tow up to 11,000 pounds, and an innovative design (Marquis, 2018; Torchinsky, 2019). The truck may be available for purchase in 2020.

Conventional automobile manufacturers in the United States are also working on electric trucks. Ford has plans to develop a plug-in hybrid pickup truck and an all-electric pickup. The earliest deliveries may be in 2020 (Capparella, 2019). There will be battery electric and hybrid versions of the model F-150. General Motors has reported that it will build an electric pickup as part of its plans for an all-electric future (Evarts, 2019c). This article also reports that GMC will build an electric SUV, that Amazon has invested about $700 million in Rivian, and that Tesla will report more on its plans for an electric truck in the summer of 2019.

Rivian and Ford have developed a technology sharing agreement, and Ford has invested $500 million in Rivian (Phelan, 2019). Phelan points out that electric trucks may have value in urban communities where the distance driven is modest so that the range of the battery is not an issue for daily work activities.

14.3 Battery Storage of Electricity

There have been continuing developments in batteries, and interest is growing in using battery storage in management of the electrical grid because of the growing investment in wind and solar generation of electricity. As of

June 2018, there are battery storage units in commercial use. The amount of energy that can be stored in a battery has units of kWh or MWh, while the flow of power into or out of the battery has units of kW or MW. In 2018, rated power was 750 MW for grid-connected battery energy storage in the USA (CSS, 2018).

In addition to grid storage, batteries are being installed in homes that have solar panels on the roof and an EV or two in the garage (Hockenos, 2019). The use of the roof for solar-generated electricity for personal home use eliminates the losses from transmission and distribution of electricity. As more people charge their EVs at home, this requires power flow to the EV, and distribution systems may need to be upgraded if the electricity comes from the utility. However, homes with solar panels and a flow of electricity from the roof to the home, and home storage, reduce the flow of electricity in the distribution system. The home with solar panels and storage batteries is an efficient option which is becoming a good economic choice.

14.4 Analysis of Co-Benefits

Many people have not considered the interrelated aspects of reducing greenhouse gas emissions and improving air quality. Both of these benefits can occur, in tandem, to varying degrees due to the reduced use of combustion methods for generating electricity. The nature of these benefits is sometimes additive, but also sometimes interactive (i.e. they build on and multiply each other). Co-benefits from reducing combustion include improved air quality and water quality, reduced greenhouse gas emissions, and reduced impacts on climate change.

When all of the benefits of transitioning to renewable energy to generate electricity are considered, the cost/benefit analysis shows many applications where one should move forward with installations of wind and solar generation of electricity (Ou et al., 2018; Scovronick et al., 2019). Because of air quality improvements, there are immediate benefits from the transition to clean power generated with wind and solar energy (Scovronick et al., 2019). Water use for wind- and solar-generated electricity is also less than that for coal-fired and nuclear power plants (Ou et al., 2018).

California officials and citizens have recognized the air quality co-benefits of reducing greenhouse gas emissions and have undertaken many initiatives designed to both improve air quality and reduce carbon emissions. Because of the challenges of finding ways to meet all of the air quality regulations, there is a major effort to electrify transportation. Some of the programs have incentives for communities that are most impacted by air pollution (CARB,

2019). Community Air Protection incentives are administered in air districts in partnership with local communities. The program includes cleaner vehicles, with priority for zero-emission projects (CARB, 2019).

Colorado officials have approved seven new bills on renewable energy, including a bill to reduce greenhouse gas emissions by 26% by 2025, 50% by 2030, and 90% by 2050, using a 2005 baseline (Kohler, 2019). Energy conservation, energy efficiency, water efficiency, extended tax credits for EVs, charging stations for EVs, and solar gardens were included in the legislation.

14.5 Wind Energy

There has been significant progress in the installation of new wind farms to generate electricity in many locations, in part because the economics around such installations have improved in the last ten years. In Kansas, for example, approximately 36% of in-state electricity generation by public utilities was from wind in 2018. This made Kansas first among the states in the percentage of electricity generated by wind, and in the top five states in terms of total wind generation (EIA, 2019). Kansas State University has signed an agreement with its public utility, Westar Energy, to obtain about 50% of its electricity from wind energy. The new agreement is projected to save the university $200,000 each year (Tidball, 2018).

References

Burgess, J. (2019). After delay, Asheville ready to get 5 electric buses on the road. *Asheville Citizen Times*, Asheville, NC; www.citizen-times.com/

Capparella, J. (2019). An all-electric Ford F-150 pickup truck is happening. *Car and Driver*; www.caranddriver.com/

CARB (2019). *Community Air Protection Incentives: 2019 Guidelines*. California Air Resources Board, Sacramento, CA.

CSS (2018). *U.S. Grid Energy Storage Fact Sheet. Pub. No. CSS15-17*. Center for Sustainable Systems, University of Michigan; css.umich.edu/

Dzikiy, P. (2019, May 30). Barcelona to receive 100+ new electric buses, replacing diesel buses. *Electrek*; electrek.co/

Eckhouse, B. (2019). The U.S. has a fleet of 300 electric buses: China has 421,000. *Bloomberg New Energy Finance*; www.bloomberg.com/

EIA (2019, May 21). *Kansas State Profile and Energy Estimates*. US Energy Information Administration; www.eia.gov/

Eisenstein, P.A. (2019). New York auto show previews Tesla's future competition with more than 40 electric vehicles on display. *CNBC*; www.cnbc.com/

Evarts, E.C. (2019a). Audi boosts plug-in plans to include 20 all-electric models by 2025. *Green Car Reports*; www.greencarreports.com/
Evarts, E.C. (2019b). BMW plans 12 all-electric models by 2025. *Green Car Reports*; www.greencarreports.com/
Evarts, E.C. (2019c). GM CEO confirms plans to build Chevy or GMC electric pickup truck. *Green Car Reports*; www.greencarreorts.com/
Evarts, E.C. (2019d). VW boosts electric car plans with more models, 22 million EVs in 10 years. *Green Car Reports*; www.greencarreports.com/
Gaton, B. (2019). EV sales surge in Europe in March quarter, as overall car sales fall. *The Driven*; thedriven.io/
Hockenos, P. (2019). In Germany, consumers embrace a shift to home batteries. *Yale Environment* 360; e360.yale.edu/
Kohler, J. (2019, May 30). Gov. Polis signs 7 bills on renewable energy, but what does that mean for Colorado's energy future? *Denver Post*; www.denverpost.com/
Marquis, E. (2018). Rivian R1T: The electric pickup with a front trunk that does 0 to 60 mph in 3 seconds. *Jalopnik*; jalopnik.com/
Ou, Y., Shi, W., Smith, S.J. et al., (2018). Estimating environmental co-benefits of U.S. low-carbon pathways using an integrated assessment model with state-level resolution. *Applied Energy* 216: 482–493.
Phelan, M. (2019, May 25). Electric pickups are coming, but does anyone really want them? Detroit Free Press; www.freep.com/
Prosser, M. (2019). China's electric buses save more diesel than all the electric cars combined. *Singularity Hub*; singularityhub.com/
Scovronick, N., Budolfson, M., Dennig, F. et al., (2019). The impact of human health co-benefits on evaluations of global climate policy. *Nature Communications*; doi.org/10.1038/s41467-019-09499-x; www.nature.com/naturecommunications/
Tidball, J. (2018). Green Energy: Wind energy agreement will provide savings, 50% of electricity needs for Kansas State University Manhattan campus. *K-State News*, November 12, 2018; www.k-state.edu/
Torchinsky, J. (2019). The Rivian pickup truck is full of fun surprises. *Jalopnik*; jalopnik.com

15

Air Quality as a Common Resource

Abstract

Air quality is a renewable common good (or common pool resource) that is both influenced by humans and in turn influences us. It has become increasingly important to understand and modify the decisions people make in shaping this relationship, in order to reduce pollution, mitigate climate change, and avoid irreparable ecological harm. In particular, there are risks for a renewable common resource being chronically overused and devolving into a public good or a club good. Further degradation can lead to collapse and destruction of the commons, with remaining resources becoming private goods. A systemic approach to these threats that uses the design principles and polycentric approach of Elinor Ostrom implicitly includes an understanding of relevant human social behaviors and decision making processes. These judgment and decision making aspects can help to draw out how and why the Ostrom approach can work to address the issue of sustainable common pool resource management. For example, the domain-specific nature of problem solving expertise supports a focus on local organizations making, modifying, and enforcing local rules to reflect changes in social, economic, and technological contexts.

15.1 Introduction

Both Japan and South Korea have been having serious issues with air pollution, but not because they are producing a lot of pollution themselves. Instead, air currents are pushing clouds of pollution to them from China (Ryall, 2017). These countries are not alone in facing this type of cross-national air pollution, with the US Environmental Protection Agency – among others – beginning to focus on the issue of transboundary air pollution (Transboundary Air Pollution, 2018). Through no fault of their own, countries are now dealing with air quality health and safety issues created by other countries, along with questions such as who should pay for the expenses associated with those issues.

Air quality is a common resource; we all benefit from and share it with everyone else. As long as there is plenty of nice, clean air for us to breathe, little thought needs to be paid to it. When something is not working, however, we need to understand enough about the system in order to fix it. Technological solutions can often get us a substantial way towards a fix, but it is almost never sufficient or stable over time. The reason technology provides only a partial solution is because human behavior is always part of the system. Even as energy production shifts to renewable resources, energy storage improves and expands, and energy use gets more efficient – at the end of the day, human decision making and behaviors will play a critical role in the success or failure of these technological innovations.

Understanding how renewable resources work as common goods, and in particular how people think and behave in such situations, is therefore important for addressing how to keep such systems working and how to fix them when challenged. You can think of it as similar to the situation of any expert who deals with a complex system (e.g. a car mechanic working on a car; a neurosurgeon working on a patient). There is the actual system (car; brain) to be dealt with, lots of technology which can aid in the process, and the expert who decides on approach and implementation. The expert's job is to fix the system when it has some kind of problem, which requires a deep understanding of what a correctly functioning system should look like. At the same time, that expert has to get the cooperation of other people (the car's owner; the patient) with regard to supportive and consistent behaviors before, during, and after making any effective changes.

This chapter is about the dynamics of air quality as a common resource system – a system that is influenced by human behaviors (e.g. carbon emissions and other pollutants), that in turn influences humans (e.g. health consequences and longer-term climate change impacts), and that can potentially be expanded or altered by technology. A central thesis of this paper is that *effectively maintaining, fixing, or even improving a system is possible only once we have a thorough understanding of how that system works.* This is not a chapter of mathematical proofs, which can be the case in economic analyses of common resources, but rather an exercise in logical analyses of what can happen when resources are overtaxed. Briefly, our focus is on what tends to happen as renewable common resources transition to other types of situations, such as public goods or club goods and even sometimes private goods. Such transitions are not inevitable, but the infrastructure around a particular resource must support the maintenance of a particular desired state in order to keep it stably in that situation.

15.2 Common Resources Can Transition Into Other Forms

Air quality is a common resource (also called a common good or a common pool resource) of a particular sort: It is a renewable common resource.

Common resources often are renewable. Technically speaking, that means that the resource increases over time, which allows people to consume a portion (i.e. fringe units) indefinitely, as long as they do not consume or damage the core portion (i.e. stock). For example, people can consume water from a spring, fish from the ocean, or lumber from a forest indefinitely, as long as the amount taken does not cut into the maintenance stock. There are also non-renewable common resources (e.g. oil, gas, and minerals), but we will put off considering these for now. Economists often characterize common resources as being relatively non-*excludable* (i.e. one cannot prevent anyone from consuming it) but *rivalrous* (i.e. one person using it prevents others from also using it). For example, I can't prevent others from fishing in the ocean, but a fish that I caught is no longer available for someone else to catch.

A major difficulty with common resources is that people are chronically tempted to take more than their fair share of the fringe units for their own use, even though doing so damages the common resources overall (Parks, et al., 2013). This can be tempting because resources taken for one's own use are entirely for one's own benefit, but the damage to the common resource overall is spread out over all the possible users of the resource. A short-term calculation of costs and benefits, therefore, can lead people to overuse the common resource – a phenomenon known as the "tragedy of the commons" (Hardin, 1968).

What happens when common resources get severely overused, with little or no effective management? One possibility is the complete collapse and destruction of the commons (e.g. overfishing leads to the extinction of the fish, or at least a reduction of the core stock to a point that use of the common resource is no longer possible). Another possibility to address a depleted common resource is to intervene by placing limits on users – circumspecting where, when, or how much of the resource can be taken. For instance, limits can be put on how many fish one person can catch, how many trees can be felled in a forest, or how much water can be taken from a river. This changes a common resource into a *club good*, which means (in economic terms) that the resource is now excludable (i.e. access is limited) and potentially non-rivalrous (i.e. rules specify how everyone can partake in the resource). There are also club goods that do not derive from renewable resources (e.g. country club memberships, cable television, and toll roads), but these are not the focus here.

Yet another possibility, and of more interest in the present context, is that some users intervene with contributions that allow the resource to be maintained. The situation is, at this point, no longer just a common pool resource but rather a *public good*. Renewable common resources become public goods (at least partially) when maintaining core stock requires investment, and that investment is provided by the community of consumers. Both public goods and common resources are non-excludable; anyone can benefit from them. But economists consider public goods, unlike common goods, to be non-rivalrous: With sufficient investment it is possible for everyone to take part in the resource consumption. For instance, a fishery can be used to restock the waters with fish, replanting can restore a forest, and water treatment plants can provide clean water. (Just like there are non-renewable common resources, which are not the focus here, there are public goods that are not derived from common resources. Examples of these include media broadcasts and information on the Internet. These public goods come into existence entirely through the efforts of the community of users, and these are also not the focus here.)

Public goods and common pool resources are often lumped together and also are frequently confused with one another. They are both theoretically and empirically distinct, though, and the differences have important consequences. Theoretically, as mentioned earlier, common pool resources and public goods are different in terms of their degree of rivalry. People can be sensitive to this difference and change how they behave in the two different types of situations (rivalrous versus non-rivalrous; Apesteguia and Maier-Rigaud, 2006). Public goods and common pool resources are also perceived to be very different in terms of the temporal sequence of events that defines each situation. For example, one summary of the two situations is that

> public goods require that citizens experience a short-term loss (of their contribution) in order to realize a long-term gain (of the good)… Common pool resources allow citizens to experience a short-term gain (by getting what they want in the early life of the resource) but also present the possibility of a long-term loss (if the resource dries up). (Parks, et al., 2013, p. 119)

More succinctly, the difference can be summed up as a "pay into" system versus a "take out" system.

Perceptions can be important, and possibly just as important as the actual structure of a system. For example, once a public good is institutionalized and in place, some people can (mis)perceive it to be a common pool resource. For instance, consider a public state university. Previous generations of residents created the university by contributing taxes (experiencing a short-term loss) in order to create the long-term gain of that public good. Subsequent generations experience the university as already existing (by either ignorance or forgetting about prior generations) and providing a stream of benefits for which they have not provided substantial investments to create. Some of those subsequent generation members may even push for tax cuts (short-term gain), even as those cuts damage the ability to provide public goods they enjoy (long-term loss). Note that this does not actually make the public good (a university, in this case) a common pool resource. Instead, this is an issue of misperception and then mistreatment based on those faulty perceptions.

In summary, when common pool resources are threatened beyond the ability for structural frameworks to resolve (Ostrom, 2010), the risk can be addressed either by active conservation (converting to a club good) or by active restoration (converting to a public good). And, of course, a mixture of both of these can be used as well. Another way to think of this, for those of an economic bent, is that problems with the optimal use of common resources can be addressed through regulations (for restoration and/or conservation) which incorporate the negative externalities to society into the costs of use by the individual consumers. The goal of this negative externality function, in an economic model, is to eliminate "deadweight loss": The difference between the optimal use of a common resource in the view of an individual versus the optimal use of a common resource in the view of the larger society (i.e. eliminating the patterns in which an individual sees more benefit and less cost to greater personal consumption, compared to a larger societal perspective).

15.3 Public and Club Goods Can Transition

As common resources get overtaxed, they may shift towards being public goods or club goods. A subsequent threat to these, however, is *free riding*. Free riding is when a person does not contribute to the investment for a public good yet still partakes of the benefits from that good – for example, poaching of fish or illegal logging. Free riding can also occur with club goods if the excludability is imperfect (e.g. stealing cable television). When free riding becomes endemic, it not only damages the stock of the resource or good, but it also can create resentment and reluctance on the part of the other users who are following the structural investment or restraint rules (e.g. Price et al., 2002).

Public and club goods that are struggling against free riders can either strengthen the infrastructure required to maintain those goods, or the situations can further evolve into *private goods*. A private good is owned by a specific individual, for use at their discretion (i.e. by themselves or by others through purchasing it from the owner), and is therefore economically both rivalrous and excludable.

Physical, moveable items (e.g. clothes, cars, tools) tend to be private goods. Non-renewable natural resources, such as fossil fuels and land mineral rights, tend to have evolved towards being private goods, even if they initially were considered common resources. (Witness the different views on land ownership between Native Americans and early European settlers in the new world, in which Europeans with a private goods view of land took advantage of Native American views of land as a common resource; e.g. Walbert, 2008.). Examples in the modern world include private parks, privately stocked fishing lakes, and bottled water. The above potential transitions are summarized in Figure 15.1.

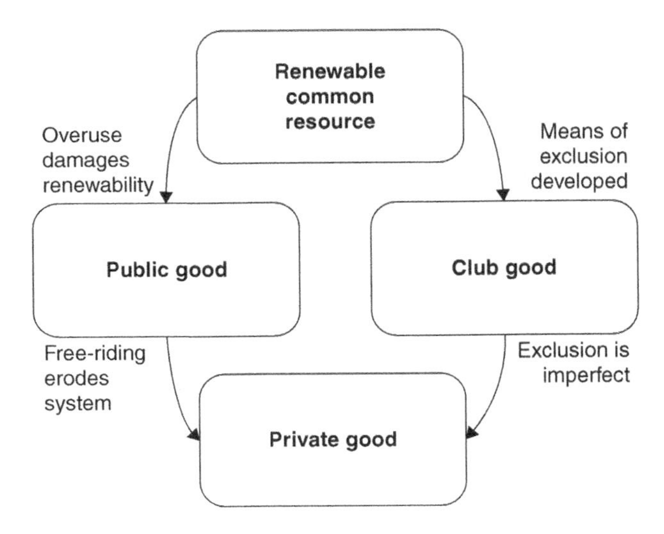

FIGURE 15.1
The potential changes of a renewable common resource (top) into either a public good or a club good, and from there into a private good.

15.4 Common Resources Can Be Maintained

Because the atmosphere is quite large, good quality air has historically been free. Clean air has been a common pool resource that has been non-excludable and only marginally (if at all) rivalrous (i.e. another person technically cannot have the breath of air that I use, but it is trivially easy for them to get air of their own). Due to these properties, clean air has remained a common resource even as other common resources (land, minerals, fish, timber, and even drinking water) have shifted towards becoming public goods, and then private goods.

Air quality has increasingly become an issue, though, pushing it out of the domain of a regular renewable common resource and into the domain of a public good. This shift is manifested by the efforts to maintain standards of air quality. Many of these efforts involve regulations to restrict air pollution (i.e. limiting carbon emissions, which are effectively overusing clean air) by factories and other industries, as the adverse health effects of air pollution have become clear (e.g. Zivin and Neidell, 2018). There are also some efforts to more proactively improve the quality of air (e.g. by planting more trees), but those efforts are challenging due to the scale of the resource (i.e. the entire atmosphere). The scale of air quality as a resource also means that the feedback loop for air quality improvements can be very slow.

In general, there have been some successes in air quality improvements once public good investments are in place to maintain an appropriate stock of the resource (i.e. air quality). For instance, major cities like Los Angeles have reduced their exceptionally high levels of air pollution, although they still have very high pollution levels (Pollack, et al., 2013). Other places have also made progress but require more extensive and active investments to reach similar outcomes (e.g. the Central San Joaquin Valley of California, including areas such as Bakersfield, Fresno, Visalia, and Merced; Air Quality Information, 2012), or new and different types of public good investments (e.g. the transnational air quality issues between China and neighboring countries).

Overuse of common pool resources can, and does, happen, but it is not inevitable. Elinor Ostrom (Ostrom, 1990, 2010; Ostrom et al., 1994) documented key features of cultural and governmental systems that can maintain stable common pool resources. Ostrom and colleagues developed a "social-ecological systems framework" to describe these features, much of which focuses on local-scale, collective self-governance and is summarized as eight design principles for sustainable common pool resource management:

1. Clear group boundaries are defined, including who is included and who is excluded and therefore unentitled.
2. Rules governing the use of common goods are matched to local needs and conditions.
3. Most or all of the people (or entities) affected by the rules can participate in modifying the rules.
4. Community members have the right to make the rules, and this right is respected by outside/higher-level authorities.
5. There is an effective monitoring system for monitoring members' behavior and enforcing rules, which is carried out by community members (or is accountable to them).
6. Rule violations are punished with graduated sanctions.
7. Disputes are addressed with low-cost and easy-to-access resolution mechanisms.
8. Responsibility for governing the common resource is organized in nested tiers, from the lowest level up to the entire interconnected system.

Additionally, some of Ostrom's later work (Dietz, et al., 2003; Ostrom, 2010; Arrow et al., 2012) stresses that an effective design needs *adaptive governance* – an ability for the system of rules to evolve and adapt over time as needed. In particular, these systems may need to adapt to reflect changes in social, economic, and technological contexts. This can include, for example, people discovering "loopholes" (ways to evade the existing governance system and personally overexploit the common resource). Much of Ostrom's work involved studying renewable common pool resources that were well managed for long periods of time, and then extrapolating the key properties of those systems. This approach –studying successful, actual systems which then inform the construction of a theoretical framework – is sometimes described as "Ostrom's law" (Fennell, 2011): "A resource arrangement that works in practice can work in theory."

Maintaining a common pool resource by adhering to Ostrom's design principles (or by any other means) is one of the few sustainable long-term solutions to the danger of overuse and eventual collapse. A currently healthy common pool resource can be prophylactically structured in this way to ensure its ongoing health. Common pool resources that have experienced some overuse or damage can be restored by implementing design principles such as Ostrom's, but this approach will only work up to a point.

The inherent nature of air quality as a property of the atmosphere has implications for how it is dealt with and for some of the issues that emerge. When air quality transitions away from being simply a renewable common resource, it tends to become a public good rather than a club good because excludability is exceedingly difficult. Outside of imagined "off-world" locations in science-fiction scenarios (where air is kept in containers of some sort), no one can prevent others from having air. The quality of that air is also generally a cumulative and diffused property, so exclusion of clean air, rather than polluted air, is difficult. (Note, however, that air filtration systems are potentially a technology that could evolve and begin to create good air quality as a club good eventually.)

Another consequence of the diffused and non-excludable nature of air quality is that it can easily and quickly become an international issue. Most considerations for common resource overuse pit individual self-interest against societal interest, with the society being a community, city, or nation-state. This level of consideration of course applies in the case of air quality (e.g. individuals polluting in contravention of regulations). Air quality conflicts also occur, though, at higher levels of organization: States or regions engaging in selfish pollution in spite of national regulations and laws; nations engaging in selfish pollution in spite of international interests. These are levels of consideration which do not exist for immobile resources (e.g. minerals and forests) but that can emerge when renewable resources are also mobile enough to easily cross national borders (e.g. water, fish, and air). The challenges of international coordination make global air quality issues particularly complex (e.g. Holzinger, 2002; Farrell and Granger Morgan, 2003).

15.5 Designing Air Quality Management

The design principles of Elinor Ostrom (1990, 2010; Dietz, et al., 2003; Ostrom et al., 1994) suggest a number of clear and concrete steps for addressing the challenge of managing air quality as a renewable common resource. These design principles also allow us to have a clearer view of some fundamental issues underlying this challenge. The most general challenge is that the relevant affected group is the entire world population, particularly when the large-scale implications of air quality degradation (such as global climate change) are recognized. Ostrom's design principles, however, make very clear that a key factor for effective resource management is local-level systems: Rules matched to local needs and conditions, local participation by people in making or modifying rules, and local monitoring and enforcement of rules. Thus, air quality is a global-level resource, but effective management needs to be driven at a local level. This is a manifestation, at the largest of scales, of the design principle that the responsibility for governing the common resource should be organized in nested tiers, from the lowest level up to the entire interconnected system.

Most of the documentation of Ostrom's design features actually implemented to address issues have been on more discrete issues, and Cox et al. (2010) note that there is a lack of use and evaluation of the principles on "larger scale cases". Although this may pose some limitations, it also is inherently consistent with design feature 8 of created nested tiers of responsibility. This nested tier system just needs to be much more extensive, addressing both nested layers of environmental problems and the nested nature of government arrangements. There have been a number of theoretical works outlining how to address air pollution and global climate change based on Ostrom's design principles as a polycentric approach (Ostrom, 2009, 2014; Dorsch and Flachsland, 2017), and there have been economic analyses (e.g. Zivin and Meidell, 2018) as well as political analyses (e.g. Chacksfield, 1993). One of the most comprehensive treatments of this topic currently is the edited volume by Jordan et al. (2018) on governing climate change.

What can be done at the highest level? Ostrom's design features indicate that international oversight of air quality as a common resource should be kept to very broad criteria of compliance and results monitoring. This restraint should be kept not because it is not possible to take a more active and involved international stance, but rather because more local involvement will generally be more effective. An obvious candidate for organization and oversight at this highest level is the United Nations, and some progress has occurred in this direction. Thus, we see the development of international regulatory efforts such as the United Nations Environment Programme and the Intergovernmental Panel on Climate Change. Ostrom's design principles have been assessed and validated, with slight elaborations by Cox et al. (2010), who heavily stress that the most effective processes are those that are very local.

The application of Ostrom's design features, though, is an inexorable push for more local control. Principles 7 and 8 make that clear: Principle 8 is that activities entailed by the other principles (appropriation, provision, monitoring, enforcement, conflict resolution, and governance activities) are organized in multiple layers of nested enterprises, and principle 7 asserts that the rights of appropriators to devise their own institutions are not challenged by external governmental authorities. Together these place the locus of power at the local level rather than at higher levels. These last two principles actually establish a context for the preceding six principles. In practice, this indicates that, whereas international groups like the United Nations can provide valuable advice on organization and oversight, the real power for action must be irrevocably delegated down to very local levels (e.g. cities and regions, such as the Los Angeles area). The United Nations can set goals and standards, but the actual governance needs to pass down to the nations, and then to regions, to cities or districts, and even to the scale of neighborhoods.

It is at the very local level that the other design features make clear sense. At the local level it is crucial to clearly establish the boundaries that define users versus non-users and that define the resource system being used versus the larger environment (principle 1). It is only at the local level that it makes sense to ensure congruence with local social and environmental conditions (principle 2), to meaningfully propose that most individuals affected by the operational rules can participate in modifying those rules (principle 3), or to establish conflict resolution mechanisms with rapid access and that operate within low-cost local arenas (principle 6). Even the monitoring of users and the status of the resource (principle 4) includes the caveat that those monitors should be accountable to the users, which is more easily and effectively done at a local level.

A clear difficulty in designing effective air quality management systems is that air is not bound to specific local users or local systems. Instead, air is inherently non-excludable. This makes boundaries (between users and non-users; between the resource and the larger environment) always blurry to some extent, and thus complicates accurate monitoring, enforcement, and sanctions, or even conflict resolution. The very nature of air quality (including greenhouse gas emissions and global climate change) requires a particularly clear and strongly structured organization of nested enterprises that can capture the issues of air quality, from the very local to the global system levels. To some extent we understand the patterns of air flow at regional, national, international, and global levels; we can in principle create a nested structure which corresponds to those patterns.

Another difficulty in designing effective air quality management is not nearly so inherent, but rather created by existing institutions. At first blush, it would seem that "users" of a resource – a fundamental unit within Ostrom's design principles – are clearly and easily defined as people. However, air quality issues are in large part influenced by users that are not individuals; they are cities, states, companies that operate over large areas, and even multinational corporations. These users are not bound by geography like a human individual, so matters once again becomes complicated and blurry when trying to apply Ostrom's design principles for accurate monitoring, enforcement, and conflict resolution with these non-person users. Consider, hypothetically, a coal-fired power plant owned by a large conglomerate corporation. As the power plant supplies electricity to the nearby city, and also pumps pollution into the air, a number of quandaries arise for Ostrom's design principles. If the winds carry the air pollution anywhere other than directly over the city using all the electricity, are the benefits proportional to the inputs of the people using that electricity? Is the monitoring of the system accountable to the users of the electricity (i.e. the local people) or the users of the coal and air (i.e. the power plant company)? If there are some users that can only be defined at higher (non-local) levels, then those users presumably must be subject to higher-level compliance and results monitoring, but do those higher-level processes need to be beholden to more local levels (per Ostrom's design principles) or beholden to even higher-level authorities? How do we get equal participation of different types of users (individuals and business/corporate entities) if these users exist at different levels of organization, yet they must in some ways all be subject to the same operational rules, monitoring, graduated sanctions for violations, and conflict resolution mechanisms?

We know *why* we need to address the challenge of air quality. Overuse (i.e. pollution, greenhouse gas emissions) threatens to devolve air quality from a sustaining, renewable common resource into a public good that requires investment to maintain. Further degradation of air quality will require even more regulations, restrictions, and restoration efforts – more costs. The bulk of this book is about *how* to address the challenge of air quality. Part of this answer involves how technology can help as a means to satisfy energy production and other needs without further damaging air quality as a renewable common resource. This includes transitions to cleaner energy production methods (e.g. solar, wind, and bioenergy), changes to energy storage and use patterns (e.g. batteries, smart grids, and electrical power management, electrification of transportation), and improvements in energy efficiency and conservation. Improving air quality is a global challenge that will not be solved with a single innovation, but rather will require an interrelated collection of new technologies, regulations, and innovations.

Another part of the answer for how to address the challenge of air quality involves innovations in the structure and organization of humans and their systems. First steps for this part of the answer include having an understanding of what that system structure is. We are living within the renewable common resource system of our atmosphere, and a general lack of good system management (e.g. excess greenhouse gas emissions) has been degrading this system (i.e. creating global climate change). We can either continue down this path, in which case air quality will transition to being a public or club good and then likely to a private good, or we can improve the management of this system – for example, by using Ostrom's design principles – and push people's behaviors towards a more sustainable system.

In conjunction with the technological advances, these may provide enough levers to stop, and even reverse, the direction of continued greenhouse gas emissions and global climate change.

15.6 Addressing Both Climate Change and Air Quality

Although climate change has been identified as the greatest global tragedy of the commons (Battersby, 2017; Paavola, 2012), the annual social cost of poor air quality was greater in 2016. The cost of disasters that were associated with climate change in 2016 was less than $1 trillion, while the cost of poor air quality was more than $3 trillion (Erickson, 2017). While things may change in the future, as it stands now, far more people die each year due to poor air quality. The human health issues associated with poor air quality provide a motivation for actions that address both challenges.

Reducing greenhouse gas emissions (i.e. addressing climate change) and improving air quality are two interrelated challenges. The Ostrom principles can be applied to address these two challenges by looking for actions that have local air quality benefits and also reduce emissions. The polycentric approach of Elinor Ostrom can also be applied to look for actions that address both issues. For example, the research and development activities to make solar- and wind-generated electricity and batteries less expensive and more competitive have been carried out using a polycentric approach. Some topics may fit better with one of these approaches than others, and some topics actually invite a mix of both. What follows in this section are some instances in which Ostrom's design principles have been primarily used, in which the polycentric approach has been primarily used, and cases that have seen a combination of both.

The Paris Agreement on Climate Change allows countries to develop their own approaches to reducing emissions. This fits with the Ostrom principles and allows for local decisions that are beneficial to local air quality as well as emissions reduction. An analysis of the Paris Agreement as it relates to the Ostrom principles was reported by Johannesson (2017), highlighting a number of strengths and weaknesses. The ability to take action at the local level is a strength because local conditions vary (e.g. because of wind and solar resources, supplies of natural gas, and current air quality). The lack of processes to discipline countries is a weakness.

Public education about the Ostrom principles and how to use them effectively is another important area. Local control necessitates that people understand the opportunities that are created by new technology and the savings that are available when it is implemented. There is likely also a need to provide education on the Ostrom principles for local, regional, and even national leaders.

Other aspects of addressing climate change and air quality may be more amenable to a polycentric approach (Paavola, 2012). This would, for example, focus on beneficial actions that are independent of one another and can be adjusted when needed as implemented to achieve environmental progress. Actions such as replacing electricity from coal-burning power plants with electricity generated with wind energy are beneficial to both improving air quality and reducing greenhouse gas emissions. When co-benefits are considered, the justification for moving forward with new installations is accepted better by decision makers (Scovronick et al, 2019). Polycentricity allows for diversity and has benefits for many organizations. Local governments, communities, non-governmental organizations, and businesses can go forward with voluntary projects that are beneficial, such as adding solar panels and charging infrastructure to parking lots. Some of these projects save money for those who participate. An example is a community group that puts solar panels on homes and reduces the cost of their electricity because of free labor.

The polycentric approach to global climate governance has been addressed, and the important features have been examined (Dorsch and Flachsland, 2017). Dosch and Flachsland (2017) address the mechanisms and features through which a polycentric approach may contribute to greater effectiveness for mitigation. Important features are: 1) self-organization, 2) an emphasis on site-specific conditions, 3) an emphasis on enabling experimentation and learning, and 4) efforts to build mutual trust to enhance cooperation (Dorsch and Flachsland, 2017). With self-organization, a group can set up its own rules and structure, making use of the Ostrom principles. There are opportunities for cost-effective actions that are site specific, and there are more of these when both local air quality and global climate change are included in the economic considerations. The site-specific considerations may be associated with the talents of the group and what they are able to do together that is beneficial. Additionally, the group may have to work effectively with the public utility, as well as the rules and regulations that exist in the community. Finally, experimentation and learning are especially important for a research group that is working to improve batteries or solar panels, but they are also important for groups that are working to improve air quality in their community. Building trust is very important in efforts to move forward with new ideas that are beneficial for a community. The implementation of a polycentric approach should be considered complementary to other approaches. One aspect of a polycentric approach is that it takes a bottom-up approach with diverse actors to address these two challenges.

Finally, topics such as voluntary green consumer behavior, such as driving electric vehicles and adopting policies such as a transition to electric buses for public transportation, are examples of actions that work well with both challenges. Carattini et al. (2017) review local social norms in addressing climate change mitigation, and they report cases and situations that appear to work cooperatively to address climate change. Social norms are able to explain much of the observed cooperative behavior, and although they can be more complex to work with technological advances, they are often very powerful. Social norms build upon individual local contributions to climate change that have global benefits.

The literature on managing the atmosphere as a global commons includes information that is beneficial in developing policy to address both challenges together (Soroos, 1995). The limits on the atmosphere as a sink for carbon dioxide, methane, and other air pollutants have received much attention because of the great impact on human health and quality of life. The feasibility of approaches to address climate change and air pollution together can be enhanced if the Ostrom principles are applied, and this includes the recognition that the methods to develop regulations and enforce them will likely vary from country to country. For instance, although actions to address air quality in a large city may be very different from the actions taken in rural areas, they both have local value and outcomes, and they both have global value in terms of reductions in greenhouse gas emissions. The Montreal Protocol is an example of a successful global agreement that has been beneficial to society because of the positive actions that have been taken to reduce emissions (Soroos, 1995).

References

Air Quality Information (2012). Retrieved from www.valleyair.org/aqinfo/aqdataidx.htm

Apesteguia, J. and Maier-Rigaud, F.P. (2006). The role of rivalry: Public goods versus common pool resources. *Journal of Conflict Resolution* 50: 646–663. doi:10.1177/0022002706290433.

Arrow, K.J., Keohane, R. O., and Levin, S.A. (2012). Elinor Ostrom: An uncommon woman for the commons. *Proceedings of the National Academy of Sciences* 109: 13,135–13,136. doi:10.1073/pnas.1210827109.

Battersby, S. (2017, January 3). Can humankind escape the tragedy of the commons? *Proceedings of the National Academy of Sciences* 114: 1, 7–10; www.pnas.org/

Carattini, S., Levin, S., and Tavoni, A. (2017). Cooperation in the climate commons. Center for Climate Change Economics and Policy Working Paper 292; www.cccep.ac.uk/

Chacksfield, A. (1993). Polycentric law versus the minimal state: the case of air pollution. *Political Notes* 76, Libertarian Alliance.

Cox, M., Arnold, G., and Tomás, S.V. (2010). A review of design principles for community-based natural resource management. *Ecology and Society* 15: 38; www.ecologyandsociety.org/vol15/iss4/art38/

Dietz, T., Ostrom, E., and Stern, P. (2003). The struggle to govern the commons, *Science* 302: 1907–1912.

Dorsch, M.J. and Flachsland, C. (2017). A polycentric approach to global climate governance. *Global Environmental Politics* 17: 2, 45–64.
Erickson, L.E. (2017). Reducing greenhouse emissions and improving air quality: Two global challenges. *Environmental Progress and Sustainable Energy* 36: 982–988.
Farrell, A.A. and Granger Morgan, M. (2003). Multilateral emission trading: heterogeneity in domestic and international common pool resource management. N. Dolšak and E. Ostrom (eds.) *The Commons in the New Millennium: Challenges and Adaptation*. MIT Press, Cambridge, MA, 169–217.
Fennell, L.A. (2011). Ostrom's Law: Property rights in the commons. *International Journal of the Commons* 5: 9–27. doi:10.18352/ijc.252.
Hardin, G. (1968). The tragedy of the commons. *Science* 162: 1243–1248. doi:10.1126/science.162.3859.1243.
Holzinger, K. (2002). Transnational common goods: Regulatory competition for environmental standards. A. Héritier (ed.) *Common Goods: Reinventing European and International Governance*. Rowman & Littlefield, Lanham, MD, 59–82.
Johannesson, E. (2017). *An Analysis of the Paris Agreement* [bachelor thesis]. Uppsala University, Uppsala, Sweden.
Jordan, A., Huitema, D., van Asselt, H., and Forster, J. (2018). *Governing Climate Change*. Cambridge University Press, Cambridge, UK.
Ostrom, E. (2009). A polycentric approach for coping with climate change. Policy Research Working Paper 5095. The World Bank, Development Economics, Office of the Senior Vice President and Chief Economist.
Ostrom, E. (2014). A polycentric approach for coping with climate change. *Annals of Economics and Finance* 15: 1, 97–134.
Ostrom, E. (2010). Beyond markets and states: polycentric governance of complex economic systems. *American Economic Review* 100: 641–72. doi:10.1257/aer.100.3.641.
Ostrom, E. (1990). *Governing the Commons: The Evolution of Institutions for Collective Action*. Cambridge University Press, Cambridge.
Ostrom, E., Walker, J., and Gardner, R. (1994). *Rules, Games, and Common Pool Resources*. University of Michigan Press, Ann Arbor.
Paavola, J. (2012). Climate change: The ultimate tragedy of the commons. Chapter 14, Cole, D.H. and Ostrom, E. (eds.), *Property in Land and Other Resources*. Lincoln Institute of Land and Policy, Cambridge, Massachusetts.
Parks, C.D., Joireman, J., and Van Lange, P.A.M. (2013). Cooperation, trust, and antagonism: How public goods are promoted. *Psychological Science in the Public Interest* 14: 119–165. doi:10.1177/1529100612474436.
Pollack, I.B., Ryerson, T.B., Trainer, M., Neuman, J.A., Roberts, J.M., and Parrish, D.D. (2013). Trends in ozone, its precursors, and related secondary oxidation products in Los Angeles, California: A synthesis of measurements from 1960 to 2010. *Journal of Geophysical Research: Atmospheres* 118: 5893–5911. doi:10.1002/jgrd.50472.
Price, M.E., Cosmides, L., and Tooby, J. (2002). Punitive sentiment as an anti-free rider psychological device. *Evolution and Human Behavior* 23: 203–231. doi:10.1016/S1090-5138(01)00093-9.
Ryall, J. (2017, May 9). True grit: clouds of Chinese dust descend on southern Japan; www.scmp.com/news/asia/east-asia/article/2093573/true-grit-clouds-chinese-dust-descend-southern-japan

Scovronick, N., Budolfson, M., Dennig, F. et al. (2019). The impact of human health co-benefits on evaluations of global climate policy. *Nature Communications* 10: 2095; www.nature.com/naturecommunications/

Soroos, M.S. (1995, May 24–28). Managing the atmosphere as a global commons. Fifth Annual Common Property Conference, International Association for the Study of Common Property, Bodo, Norway.

Transboundary Air Pollution (2018, January 28). Retrieved from www.epa.gov/international-cooperation/transboundary-air-pollution

Walbert, D. (2008) Who owns the land? *North Carolina: A Digital History*. Digital Textbook. Accessed (23 Jan 2018); www.learnnc.org/lp/editions/nchist-colonial/2027

Zivin, J.G. and Neidell, M. (2018) Air pollution's hidden impacts. *Science* 359. doi:10.1126/science.aap7711.

16

Conclusions

Abstract

We are making great progress with the important developments that are needed with respect to the interrelated challenges of reducing greenhouse gas emissions and improving air quality. Between 2009 and 2019, wind and solar energy technology has steadily and significantly improved, while prices for that technology have at the same time continued to decrease and become competitive with other energy sources. Batteries for electric vehicles and storage of energy for the electrical grid have similarly improved in quality and decreased in price, enabling a transition to their use for transportation and electric power management. As these trends continue, the future is looking good. Research and development is continuing on solar panels, batteries, wind turbines, and electric vehicles because there is global competition to be the leader in these areas, with the product of choice in the marketplace.

16.1 Introduction

From our current vantage point, we can look back and see significant changes over the past ten years in addressing greenhouse gas emissions and air quality, including the technologies associated with these issues: Renewable energy technology, battery and electric vehicle technologies, and gradual improvements of the electrical grid and its management. There have also been societal changes in policies, economics, and our understanding of air quality as a renewable common good resource. Some of these developments have taken longer than others; some of them have changed just in the time that the authors have been writing this book. This chapter provides some information on the progress and trends that relate to the topics of the book, along with some conclusions with respect to the main content and subjects.

In the Introduction we compared the steps we need to take with regard to reducing greenhouse gas emissions and improving global air quality to setting out on a new and still dimly lit path. No one likes to walk an unfamiliar path in the dark. Hopefully this book has helped to provide illumination across the diverse, rapidly changing, and exciting aspects of this new path. It turns out that this path is more like several trails that are all going in the same general direction, often intersecting, so we have used a broad and diffuse light. There are things going on in terms of wind, solar, and other forms of renewable energy generation. There are ongoing advances in battery storage and the electrification of transportation. These changes are spurring development of the smart grid, improved electric power management, and responsive government policies. And underlying all of these developments are people: People working out the economics of these new pathways in entire countries, in big cities, and in agriculture. This is even leading towards a more advanced understanding of our relationship with our global atmosphere and its climate.

16.2 Wind and Solar Energy

Within the broad goal of generating electricity without combustion and carbon emissions, there has been tremendous recent progress in the generation of electricity from wind and solar energy. Wind is now the low-cost alternative for generation of electricity in some locations, and the economics of solar-generated electricity has become increasingly commercially viable (IRENA, 2019). As of 2019, wind and solar energy are the lowest-cost sources of new power generation in most parts of the world. Moving into the future, new wind and solar installations will begin to produce electricity that costs less than the operating cost of existing coal-fired power plants in many cases. IRENA's analysis of price trends for the weighted average cost of electricity projects $0.049/kWh for wind and $0.055/kWh for solar photovoltaics (PV) in 2020 (IRENA, 2019). These values may be compared to $0.047/kWh for hydro, $0.056/kWh for wind, and $0.085 for solar PV in 2018 (IRENA, 2019). The April 2019 World Economic Outlook (Bogmans, 2019) summarizes the decreases in costs of wind- and solar-generated electricity from 2009 to 2017. The levelized cost of electricity from solar PV has decreased by 76%, and the cost of electricity from wind has decreased by 34% over this time period.

Solar, wind, and hydro generation of electricity have each become competitive, and each technology has the great advantage of producing electricity without air pollution and carbon emissions. A significant fraction of new generating capacity that was installed in 2018 was renewable. Approximately 45 GW of new wind generating capacity was installed in 2018 globally (IRENA, 2019). These new installations are progressively changing the shape of the energy production sector. In Kansas, for example, approximately 36% of electricity generation by public utilities was from wind in 2018 (EIA, 2019).

16.3 Batteries and Electric Power Management

As the fraction of electricity produced from wind and solar power generation increases, the challenge of effectively using all of the electricity that is generated must be addressed. The issues of demand management (Chapters 7–9) include a constellation of improvements and additions to the electrical grid, producing the so-called smart grid. This smart grid is necessary for effective power management, either within a large system such as a city or in smaller "off-grid" systems that must be efficient and responsive on their own. The fluctuations in power supply from wind and solar energy generation make battery storage a key part of electric power management. The decrease in battery costs is, of course, beneficial here, as it allows for more capacity and lower costs to address issues in storing energy for grid management of supply and demand. Larger, more efficient, and less expensive batteries also have implications for electric vehicles (EVs), both in terms of their appeal and their potential for interacting with the smart grid.

The prices of batteries, both for EVs and electrical energy storage, have decreased from about $1000/kWh in 2010 to less than $200/kWh in 2018. Because of rapidly falling battery prices, new developments, and new models, consumers have many different models of EVs to choose from. New global EV sales reached more than 2 million electric cars for the first time in 2018 (Loveday, 2019). The prices of batteries are expected to continue to decrease, and batteries are becoming a logical choice to store energy when the electrical grid has energy that needs to be stored. As research on batteries continues, we can expect additional progress in materials, energy density, cost, and durability.

16.4 Electrification of Transportation

The electrification of transportation is very quickly developing into more than just electric cars. Since transportation is one of the most significant sources of air pollution, the progress in developing electric cars, buses, trucks, and bicycles holds great value for society. Those benefits, relative to conventional transportation, multiply if the vehicle batteries are charged with electricity from renewable energy.

The transition to electric buses is in progress in 2019, and it is already clear that it is able to provide great benefits not only to people who ride electric buses but to all the people around them who experience better air quality. The decisions that are being made to transition to electric buses are based on reduced air pollution, a quieter ride, and reduced maintenance costs (RTA, 2019). For every 1000 electric buses, about 500 barrels of diesel fuel are not combusted each day (RTA, 2019). We can see this in China, for instance, where the electrification of bus transportation has progressed quite a bit. In Shenzhen, all 16,359 buses are now electric, and BYD reached 50,000 electric buses manufactured in January 2019. With 385,000 electric buses on the road in 2019, 17% of China's bus fleet is powered by electricity. In addition to China, more than 1600 electric buses are present in Europe in many cities, including London, Paris, Oslo, and Copenhagen in 2019. In Latin America, there are electric buses on order or in service in Argentina, Brazil, Chile, Colombia, Costa Rica, and Ecuador (Scriven, 2019). The issues of range and charging time have been addressed in many applications. For example, some buses in London are used all day, racking up 180 miles traveled on a single charge, and then they are recharged at night (RTA, 2019).

In 2018, California became the first state in the USA to require transit agencies to stop purchasing buses powered by fossil fuels by 2029 (Fehrenbacher, 2019). Fehrenbacher (2019) points out that there is interest in having electric school buses because of the impact of diesel fumes on health. Proterra and Daimler are working together on the development of electric school buses (Fehrenbacher, 2019).

Another nascent but exciting new market is for electric trucks. This can be particularly well suited for company fleets of delivery vans that have set routes with known mileages and energy needs. The fleet market for EVs is expected to be large. Fleets are typically purchased based on established needs and economic considerations, so the decreases in battery prices (and increases in capacity) will progressively make small electric fleet vehicles the lowest-cost alternative very soon. There will be significant benefits to urban air quality as fleet vehicles become EVs.

In many parts of the world, there is a need for charging infrastructure for EVs, and this topic was covered in an earlier book (Erickson et al., 2017). Because of the importance of having EVs connected to charging stations for management of supply and demand of renewable electricity in the grid, more charging infrastructure is recommended.

16.5 Technology in Context

The advances in technology reviewed in this book are wide-ranging and stunning. There have been rapid advances in renewable wind and solar energy, in battery storage, in energy management and smart grid technology, and, of course, across an array of different EVs. These technological improvements, including in efficiency and production techniques, have inexorably pushed the basic economics towards the adoption of these improvements across all applicable areas of life. We have to keep in mind, though, that there are also a number of social factors that play strong roles in the adoption of these new technologies.

At a broad level, the social role of new technology adoption plays out in terms of government policies. Policies can be made at the global, national, regional, and local levels. Cumulatively, they can not only influence the actual economics of different technologies but also the perceptions of the economic situation. Policies and actual locations interact; the resources and demands for energy are different in big cities, in rural areas (including for agriculture), in developing versus developed countries. Policies and processes for reducing greenhouse gas emissions and improving air quality need to reflect local conditions and be under local control in order to achieve that. Elinor Ostrom's principles for managing common goods provide explicit guidelines like this – and others – for structuring the social situation for maintaining the renewable common good that is our atmosphere.

16.6 The Future of Reducing Emissions and Improving Air Quality

In 2015, the estimated number of early deaths due to poor air quality in China was reported to be 1,600,000 (RTA, 2019). When the article 'Reducing Greenhouse Gas Emissions and Improving Air Quality: Two Global Challenges' was published in 2017 (Erickson, 2017), the connection between the two actions was not as widely appreciated as it is now in 2019. Because many of the actions that need to be taken to improve urban air quality are now cost effective to do, the years ahead will see more and more actions that will reduce emissions and improve air quality. Indeed, it is already documented that the benefits of actions that improve air quality often exceed the costs of doing those actions (Gardiner, 2019). This is particularly the case for the many large cities of the world that do not meet the recommended air quality standards for good health. Replacing coal-fired power plants with electricity from renewable energy and transitioning to EVs for transportation are now possible and these transitions can be accomplished cost-effectively.

Where do things go from here? In certain respects, the current trends will continue: Renewable energy, EVs, batteries, and the electrical grid will all keep getting progressively better. People will gradually adapt. There are a few things, however, that are not as immediately obvious.

One of the significant developments that needs to be stressed is the great decreases in the costs of solar panels and batteries. Both of these have become much more affordable, not only changing the economics in existing infrastructures, but also fundamentally creating opportunities that previously did not exist. For instance, people who live where there is no electrical grid can get stand-alone electrical systems because the cost has been reduced by about 80%. The opportunity to have an off-grid electrical system is now possible for many more people. Developments such as these additionally feed into the Sustainable Development Goals, and they are good examples of what can to be done to improve global quality of life. Electricity can be used for cooking, replacing solid fuel open fires, which impacts the health and safety of the cooks as well as improving air quality.

Predicting the future is always a dicey thing. It is fairly safe to say, though, that there will be additional progress towards lower prices and improved efficiency of solar panels, and lower prices and greater energy density of batteries. EVs will become more popular because of lower costs, better features, and more models. As EVs become common enough that they are perceived as normal choices rather than novelties, we can expect an inflection point – EVs will become the status quo (Brase, 2019). We predict that roughly 50% of the readers who read this sentence will drive or ride in an EV (bike, car, bus, truck) sometime during the next five years.

With the progressive electrification of transportation will come more focus on the sources of electricity, which can combine with concerns about air quality and greenhouse gas emissions. Other states, cities, and countries will follow the leaders (California, China, and Norway) and make a greater effort to improve air quality by transitioning to renewable energy and zero-emission vehicles. Development of the smart grid will continue and greater use will be made of real-time or time-of-use prices. We also expect that there will continue to be global competition in manufacturing operations for solar panels, solar lanterns, batteries, wind turbines, and all types of EVs (electric bikes, cars, buses, and trucks). This global competition will lead to further developments that will be beneficial to society.

References

Bogmans, C. (2019, April). Falling costs make wind, solar more affordable. World Economic Outlook, International Monetary Fund; blogs.imf.org/

Brase, G.L. (2019). What would it take to get you into an electric car? Consumer perceptions and decision making about electric vehicles. *The Journal of Psychology: Interdisciplinary and Applied* 153: 214–236. doi:10.1080/00223980.2018.1511515.

EIA (2019, May 21). *Kansas state profile and energy estimates*. US Energy Information Administration; www.eia.gov/

Erickson, L.E. (2017). Reducing greenhouse gas emissions and improving air quality; Two global challenges. *Environmental Progress and Sustainable Energy* 36: 982–988.

Erickson, L.E., Robinson, J., Brase, G., and Cutsor, J. (2017). *Solar Powered Charging Infrastructure for Electric Vehicles: A Sustainable Development*. CRC Press, Boca Raton, Florida.

Fehrenbacher, K. (2019, March 25). Electric buses and trucks charge ahead. *Green Biz.*; www.greenbiz.com/

Gardiner, B. (2019) *Choked: Life and Breath in the Age of Air Pollution*. University of Chicago Press.

IRENA (2019). *Renewable Power Generations Costs in 2018*. International Renewable Energy Agency, Abu Dhabi; www.irena.org/

Loveday, S. (2019). Final Update: Monthly Plug-in EV Sales Scorecard: May 2019. *InsideEVs*; insideevs.com/

RTA (2019, May 14). All aboard the electric bus – Modern public transport powered by electricity is coming back. *Rapid Transition Alliance*; www.rapidtransition.org/

Scriven, R. (2019, March 13). LATAM electric bus market is strong, but may fail to reach its potential. *Intelligent Transport*; www.intelligenttransport.com/

Index

Page numbers in **bold** indicate tables and in *italic* indicate figures.

For Product Safety Concerns and Information please contact our EU representative GPSR@taylorandfrancis.com Taylor & Francis Verlag GmbH, Kaufingerstraße 24, 80331 München, Germany

Printed and bound by CPI Group (UK) Ltd, Croydon, CR0 4YY
08/06/2025
01897007-0013